MW00913779

INDOLOGY

2003

NARSIDASS

HERS

उर्ध्वतिर्यग्भ्यामं

VERTICALLY AND CROSSWISE

Applications of the Vedic Mathematics Sutra

India's Scientific Heritage

General Editor: Dr L M Singhvi

6

Editorial Panel

Abhijit Das

Andrew Nicholas

Ashutosh Urs Strobel

B D Kulkarni

David Frawley

David Pingee

James T Glover

Jeremy Pickles

Kenneth R Williams

K V Sarma

M A Dhaky

Mark Gaskell

Navaratna S Rajaram

P K Srivathsa

R K Tiwari

Rajiv Malhotra

Sambhaji Narayan Bhavsar

Subhash Kak

Toke Lindegaard Knudsen

V V Bedekar

Vithal Nadkarni

W Bradstreet Stewart

उर्ध्वतिर्यग्भ्यामं

VERTICALLY AND CROSSWISE

Applications of the Vedic Mathematics Sutra

A.P. NICHOLAS
K.R. WILLIAMS
J. PICKLES

Foreword by
L.M. SINGHVI
Formerly High Commissioner for India in the UK

MOTILAL BANARSIDASS PUBLISHERS
PRIVATE LIMITED • DELHI

First Indian Edition: Delhi, 2003

First Edition: 1984

(First Published by Inspiration Books)

ISBN: 81-208-1960-8 (Cloth)
ISBN: 81-208-1982-9 (Paper)

Also available at:

MOTILAL BANARSIDASS

41 U.A. Bungalow Road, Jawahar Nagar, Delhi 110 007
8 Mahalaxmi Chamber, 22 Bhulabhai Desai Road, Mumbai 400 026
120 Royapettah High Road, Mylapore, Chennai 600 004
236, 9th Main III Block, Jayanagar, Bangalore 560 011
Sanas Plaza, 1302 Baji Rao Road, Pune 411 002
8 Camac Street, Kolkata 700 017
Ashok Rajpath, Patna 800 004
Chowk, Varanasi 221 001

Printed in India
BY JAINENDRA PRAKASH JAIN AT SHRI JAINENDRA PRESS,
A-45 NARAINA, PHASE-I, NEW DELHI 110 028
AND PUBLISHED BY NARENDRA PRAKASH JAIN FOR
MOTILAL BANARSIDASS PUBLISHERS PRIVATE LIMITED,
BUNGALOW ROAD, DELHI 110 007

FOREWORD

Through trackless centuries of Indian history, Mathematics has always occupied the pride of place in India's scientific heritage. The poet aptly proclaims the primacy of the Science of Mathematics in a vivid metaphor:

यथा शिखा मयूराणां, नागानां मणयो यथा।
तद्वद् वेदांगशास्त्राणाम् गणितं मूर्ध्नि स्थितम्॥

—वेदांग ज्योतिष*

"Like the crest of the peacock, like the gem on the head of a snake, so is mathematics at the head of all knowledge."

Mathematics is universally regarded as the science of all sciences and "the priestess of definiteness and clarity". J.F. Herbert acknowledges that "everything that the greatest minds of all times have accomplished towards the comprehension of forms by means of concepts is gathered into one great science, Mathema-tics". In India's intellectual history and no less in the intellectual history of other civilisations, Mathematics stands forth as that which unites and mediates between Man and Nature, inner and outer world, thought and perception.

Indian Mathematics belongs not only to an hoary antiquity but is a living discipline with a potential for manifold modern applications. It takes its inspiration from the pioneering, though unfinished work of the late Bharati Krishna Tirthaji, a former Sankaracharya of Puri of revered memory who reconstructed a unique system on the basis of ancient Indian tradition of mathematics. British teachers have prepared textbooks of Vedic Mathematics for British Schools. Vedic mathematics is thus a bridge across centuries, civilisations, linguistic barriers and national frontiers.

Vedic mathematics is not only a sophisticated pedagogic and research tool but also an introduction to an ancient civilisation. It takes us back to many millennia of India's mathematical heritage. Rooted in the ancient Vedic sources which heralded the dawn of human history and illumined by their erudite exegesis, India's intellectual, scientific and aesthetic vitality blossomed and triumphed not only in philosophy, physics, astronomy, ecology and performing arts but also in geometry, algebra and arithmetic. Indian mathematicians gave the world the numerals now in universal use. The crowning glory of Indian mathematics was the invention of zero and the introduction of decimal notation without which mathematics as a scientific discipline could not have made much headway. It is noteworthy that the ancient Greeks and Romans did not have

---

* Lagadha, Verse 35

the decimal notation and, therefore, did not make much progress in the numerical sciences. The Arabs first learnt the decimal notation from Indians and introduced it into Europe. The renowned Arabic scholar, Alberuni or Abu Raihan, who was born in 973 A.D. and travelled to India, testified that the Indian attainments in mathèmatics were unrivalled and unsurpassed. In keeping with that ingrained tradition of mathematics in India, S. Ramanujan, "the man who knew infinity", the genius who was one of the greatest mathematicians of our time and the mystic for whom "a mathematical equation had a meaning because it expressed a thought of God", blazed new mathematical trails in Cambridge University in the second decade of the twentieth century even though he did not himself possess a university degree.

I do not wish to claim for Vedic Mathematics as we know it today the status of a discipline which has perfect answers to every problem. I do however question those who mindlessly deride the very idea and nomenclature of Vedic mathematics and regard it as an anathema. They are obviously affiliated to ideological prejudice and their ignorance is matched only by their arrogance. Their mindset were bequeathed to them by Macaulay who knew next to nothing of India's scientific and cultural heritage. They suffer from an incurable lack of self-esteem coupled with an irrational and obscurantist unwillingness to celebrate the glory of Indian achievements in the disciplines of mathematics, astronomy, architecture, town planning, physics, philosophy, metaphysics, metallurgy, botany and medicine. They are as conceited and dogmatic in rejecting Vedic Mathematics as those who naively attribute every single invention and discovery in human history to our ancestors of antiquity. Let us reinstate reasons as well as intuition and let us give a fair chance to the valuable insights of the past. Let us use that precious knowledge as a building block. To the detractors of Vedic Mathematics I would like to make a plea for sanity, objectivity and balance. They do not have to abuse or disown the past in order to praise the present.

Dr. L.M. Singhvi
Formerly High Commissioner for India in the UK

ACKNOWLEDGEMENT

We would like to thank Mark Riley for checking the answers to, and producing the exercises for, Chapters 2 - 7 and 12 - 16.

PREFACE TO SECOND EDITION

Numerous corrections have been made to the first edition; there are also a few minor changes of text for the sake of greater clarity. The examples are as before.

APN 1998

In celebration of the centenary of the birth of
SRI BHARATI KRISHNA TIRTHAJI (1884-1960)

PREFACE

PART A: Introductory

Vedic Mathematics offers a fresh and highly efficient approach to mathematics covering a wide range—this book starts with elementary multiplication and concludes with a relatively advanced topic, the solution of non-linear partial differential equations. But the vedic scheme is not simply a collection of rapid methods: it is a system, a unified approach, which can be swiftly learnt.

Although primarily intended for mental working, this approach has considerable potential for automatic computation. Good tools yield best results in the hands of skilled craftsmen. The computer is a powerful tool, and to be well-used it requires application of the best available relevant skills.

What is the essence of the system? Bharati Krishna Tirthaji, who rediscovered it earlier this century, tells us it is based on sixteen sutras. A sutra is a terse statement of an important point or principle.

Ultimately the essential nature of the approach is to be discovered by using it. Nevertheless, certain points can be made. In arithmetic, it uses positional notation. In particular, it uses the decimal system, which is regarded as using a pattern made of nine elements, the numbers 1 to 9, together with the zero, which acts as a spacing element. Something resembling the attitude in question is found in our use of telephone numbers, in which quantitative considerations are of the least importance, and the emphasis is on the pattern. Thus the number 243 is treated as a pattern of 2 followed by 4 followed by 3, a qualitative entity, rather than as the quantitative entity 'two hundred and forty three' (which would signify nothing to a telephone exchange).

The fact that 'one' stands for a single unit, tens unit, one hundred, one thousand, etc. in the decimal system makes in a sense no difference to its 'quality' as the number one—it retains its oneness. The same of course is true of all the other elements up to nine in that system. Should the base be different from ten, this same fundamental fact applies to the elements needed for that base.

In the vedic scheme, each number is also not simply considered in itself, but it is, in general, seen in relation to a base. Thus '8' brings to mind a deficiency of 2 from the base of

10, '7' that of 3 from the same base, etc. As some early examples in Chapter 1 show, this simple device can lighten the task enormously at times.

Vedic Mathematics extensively exploits the properties of numbers in very practical applications, particularly in the field of computation. It makes available a whole range of methods ideally suited to these properties. Owing to the nature of the sutra, the book *Vertically and Crosswise* is concerned mainly with general methods. But the vedic approach has also its special methods—indeed rather more of them than conventional mathematics. The situation is not unlike that of a skilled craftsman who has a whole array of tools to draw from and selects whatever is most suitable for the work in hand. This large flexibility of method finds itself reflected in the mind when approaching problems from the vedic viewpoint. As a doctor in biochemistry remarked during a presentation, 'My mind works this way'. This indeed is the great benefit of the approach, particularly when taught to younger students. Presented by a skilful teacher, its simplicity and ease readily shine forth, and one is left wondering at the reason, or lack of it, for most of our usual methods. By use of aphorisms (or 'sutras') easy to memorise, it is perfectly adapted to oral teaching and mental calculation. The only additional need is plenty of simple practice.

The plan of the book *Vertically and Crosswise* is essentially as follows:

1. It is concerned with calculation, and the evaluation of functions.
2. It deals with the solution of equations, beginning with simultaneous linear equations, and moving on to algebraic, transcendental, and differential equations.

These days there is considerable interest in the use of bases other than ten. That algebra is a generalisation of arithmetic is widely understood. What is not always appreciated is that the polynomial acts as a generalisation of positional notation, the base of 10 or 2 or whatever being replaced by a base of x, or some other letter. The student may be held back by various hindrances to this appreciation. One is simply overlooking that $x^0 = 1$. Another is that digits recorded in positional notation are, by custom, all positive, whereas the constants of a polynomial can be positive or negative. The vedic scheme uses both positive and negative digits however, which deals with this point (see Chapter 1). A third difficulty is that, in positional notation, it is not customary to use digits exceeding the base—rather than this we carry figures into the next place. But this limitation is not essential. It is quite feasible to have numbers exceeding the base, in any given position, and the vedic plan allows for and makes use of this.

Once the important connection between algebra and arithmetic is established, that a polynomial can be seen as a positional notation with the base unspecified, an important question follows: can we, in that case, have an efficient system of computation applicable to both arithmetic and algebra? That the vedic scheme makes provision for this is shown in this book. In consequence, on having learned arithmetic methods, very little extra is required to learn algebraic ones. This is something which a study of the 'vertically and crosswise' sutra demonstrates. The ramifications are considerable—to the extent that the very last chapter of this book rests essentially on methods expounded in Chapter 1.

PART B: Historical and Cultural Aspects

At this point the reader might welcome some historical background. The vedic tradition was originally oral. Memory was aided by versification, and by terse statements of important points, called 'sutras'. When the vedic period began is uncertain, but according to modern scholars the Vedas began to be written down about 1600 or 1700 BC. There are four of them: Rig-, Sama-, Yajur-, and Atharva-veda. In the course of time, no doubt influenced by the numerous invasions of India, much of the vedic tradition fell into disuse. Then in the 19th century scholars took renewed interest in the Vedas. It is recounted in the *author's preface to 'Vedic Mathematics'* how they could make no sense of the mathematical portion of the Vedas (an appendix to the Atharvaveda). Then Sri Bharati Krishna Tirthaji, after lengthy and careful investigation, produced a reconstruction of the ancient mathematical system based on sixteen sutras, together with a number of sub-sutras. He wrote sixteen volumes on the subject, all of which were subsequently lost. Once the loss was confirmed he decided to rewrite all sixteen volumes, and began by writing 'Vedic Mathematics', intended as an introductory volume. Poor health and then death prevented him from writing any further volumes.

Intriguing historical questions are posed by Tirthaji's reconstruction. Not least is the problem of finding the historical evidence used for the reconstruction. His introduction to 'Vedic Mathematics' tells us practically all we know about how the system was reconstructed. Yet the references he gives do not seem to be available, or else those versions which are available do not include the material he refers to. Thus, until further light is thrown on the matter, perhaps the minds of historians ought to be open to a number of possibilities concerning the status of this reconstruction, ranging from its being a work inspired by the vedic tradition, to its being an accurate account of a system used in vedic times. At times, Sri Bharati Krishna Tirthaji seems to adopt the latter view, and it is as well to respect the views of such an outstanding man, bearing in mind his considerable learning in Sanskrit, in mathematics, and in the vedic tradition (as well as in other areas), and also his undoubted integrity. But his main interest seems to be in what he considers the Vedas should be: an all-round system of knowledge.

Exposure to this approach eventually shows us that we are dealing with a new way of thinking. One can learn to speak a little French, but think in English. To really enter the spirit of a language one needs to think in terms of that language. Similarly, if he wishes to gain most from this system, the western-trained mathematician is cautioned against trying to fit this system into the mental framework he already has, which is what most western training encourages us to do. If the methods are practised with a neutral attitude of mind, allowing the system to speak for itself, then the possibility is opened of entering the spirit of this approach. Otherwise there is a risk of simply acquiring a few techniques, and not really gaining any overall sense of the approach. This caution apart, the fresh insights offered by this plan should be welcome to western mathematicians: there is much to be said for having two strings to one's bow.

With the rapid growth of communications, and the spread of science and technology, the differences between East and West are, in some ways, not so great as they were. One such difference which remains, however, concerns modes of address. As well as being an outstanding scholar, Tirthaji was a very saintly man. Unwillingly he found himself thrust into the position of Shankaracharya, one of four spiritual heads of Hindu India. As such he became known in India as 'Sri Jagadguru Bharati Krishna Tirthaji, Shankaracharya of Puri' or 'The Shankaracharya of Puri', or 'Sri Bharati Krishna Tirthaji'. In the West he is known simply as 'Tirthaji'. Different perceptions lie behind these modes of address. The western viewpoint is, perhaps, that to be acknowledged as a member of the human race is honour enough, and that further titles are superfluous.

The eastern world, by contrast, still pays considerable respect to a man's office. These are cultural differences, and people of each culture need to respect the existence of a different point of view. May India continue to uphold her ancient and colourful tradition, despite the world-wide tendency towards a drab monoculture!

The present work has been put together with more haste than the authors would have liked. It is hoped readers will take more interest in what it has to offer than any defects it may contain. Nevertheless, we would appreciate having our attention drawn to any errors or omissions.

This book is intended to commemorate the birth of Sri Bharati Krishna Tirthaji. As such it is perhaps appropriate to conclude with a few remarks about the man and his ideals. He believed in the ancient tradition of all-round spiritual and cultural harmony, and his ambition for humanity was a world-wide cultural and spiritual renewal. An excellent brief account of the man, his life and works, is given by Mrs Manjula Trivedi in her preface to 'Vedic Mathematics'. She looked after him during the last few years of his life, and is now in charge of the foundation he set up in Nagpur in 1953, the 'Shri Vishwapunarniman Sanga' (World Reconstruction Association). It was founded to act as a vehicle for world renewal.

A. P. Nicholas
K. R. Williams
J. Pickles

London, July 1984

A- A DESCRIPTIVE PREFATORY NOTE ON THE ASTOUNDING
WONDERS OF ANCIENT INDIAN VEDIC MATHEMATICS

1. In the course of our discourses on manifold and multifarious subjects (spiritual, metaphysical, philosophical, psychic, psychological, ethical, educational, scientific, mathematical, historical, political, economic, social etc., etc., from time to time and from place to place during the last five decades and more, we have been repeatedly pointing out that the Vedas (the most ancient Indian scriptures, nay, the oldest "Religious" scriptures of the whole world) claim to deal with all branches of learning (spiritual and temporal) and to give the earnest seeker after knowledge all the requisite instructions and guidance in full detail and on scientifically—nay, mathematically— accurate lines in them all and so on.

2. The very word "Veda" has this derivational meaning, i.e. the fountain-head and illimitable store-house of all knowledge. This derivation, in effect, means, connotes and implies that the Vedas should contain within themselves all the knowledge needed by mankind relating not only to the so-called 'spiritual' (or other-worldly) matters but also to those usually described as purely "secular", "temporal", or "worldly"; and also to the means required by humanity as such for the achievement of all-round, complete and perfect success in all conceivable directions and that there can be no adjectival or restrictive epithet calculated (or tending) to limit that knowledge down in any sphere, any direction or any respect whatsoever.

3. In other words, it connotes and implies that our ancient Indian Vedic lore should be all-round complete and perfect and able to throw the fullest necessary light on all matters which any aspiring seeker after knowledge can possibly seek to be enlightened on.

4. It is thus in the fitness of things that the Vedas include: (i) *Ayurveda* (anatomy, physiology, hygiene, sanitary science, medical science, surgery etc., etc.,) not for the purpose of achieving perfect health and strength in the after-death future but in order to attain them *here and now* in our present physical bodies; (ii) *Dhanurveda* (archery and other military sciences) not for fighting with one another after our transportation to heaven but in order to quell and subdue all invaders from abroad and all insurgents from within; (iii) *Gandharva Veda* (the science and art of music) and (iv) *Sthapatya Veda* (engineering, architecture etc., and all branches of mathematics in general). All these subjects, be it noted, are inherent parts of the Vedas i.e. are reckoned as "spiritual" studies and catered for as such therein.

5. Similar is the case with regard to the *Vedangas* (i.e. grammar, prosody, astronomy, lexicography etc., etc.) which, according to the Indian cultural perceptions, are also inherent parts and subjects of *Vedic* (i.e. *Religious*) study.

6. As a direct and unshirkable consequence of this analytical and grammatical study of the real connotation and full implications of the word "Veda" and owing to various other historical causes of a personal character (into details of which we need not now enter),

we have been from our very early childhood, most earnestly and actively striving to study the Vedas critically from this stand-point and to realise and prove to ourselves (and to others) the correctness (or otherwise) of the derivative meaning in question.

7. There were, too, certain personal historical reasons why in our quest for the discovering of all learning in all its departments, branches, sub-branches etc., in the Vedas, our gaze was riveted mainly on ethics, psychology and metaphysics on the one hand and on the "positive" sciences and especially mathematics on the other.

8. And the contemptuous or, at best patronising attitude adopted by some so-called Orientalists, Indologists, antiquarians, research-scholars etc., who condemned, or light-heartedly, nay; irresponsibly, frivolously and flippantly dismissed, several abstruse-looking and recondite parts of the Vedas as "sheer-nonsense"—or as "infant-humanity's prattle", and so on, merely added fuel to the fire (so to speak) and further confirmed and strengthened our resolute determination to unravel the too-long hidden mysteries of philosophy and science contained in India's Vedic lore, with the consequence that, after eight years of concentrated contemplation in forest-solitude, we were at long last able to recover the long lost keys which alone could unlock the portals thereof.

9. And we were agreeably astonished and intensely gratified to find that exceedingly tough mathematical problems (which the mathematically most advanced present-day Western scientific world had spent huge lots of time, energy and money on and which even now it solves with the utmost difficulty and after vast labour and involving large numbers of difficult, tedious and cumbersome "steps" of working) can be easily and readily solved with the help of these ultra-easy Vedic Sutras (or mathematical aphorisms) contained in the Parishishta (the Appendix-portion) of the ATHARVAVEDA in a few simple steps and by methods which can be conscientiously described as mere "mental arithmetic".

10. Ever since (i.e. since several decades ago), we have been carrying on an incessant and strenuous campaign for the India-wide diffusion of all this scientific knowledge, by means of lectures, blackboard-demonstrations, regular classes and so on in schools, colleges, universities etc., all over the country and have been astounding our audiences everywhere with the wonder and marvels not to say, miracles of Indian Vedic Mathematics.

11. We were thus at last enabled to succeed in attracting the more than passing attention of the authorities of several Indian universities to this subject. And, in 1952, the Nagpur University not merely had a few lectures and blackboard-demonstrations given but also arranged for our holding regular classes in Vedic Mathematics (in the University's Convocation Hall) for the benefit of all in general and especially of the University and college professors of mathematics, physics etc.

12. And, consequently, the educationists and the cream of the English educated section of the people including the highest officials (e.g. the high-court judges, the ministers etc.,) and the general public as such were all highly impressed; nay, thrilled, wonder-struck and flabbergasted! and not only the newspapers but even the University's official reports described the tremendous sensation caused thereby in superlatively eulogistic terms; and the papers began to refer to us as "the Octogenarian Jagadguru Shankaracharya who had taken Nagpur by storm with his Vedic Mathematics", and so on!

13. It is manifestly impossible, in the course of a short note (in the nature of a "trailer"), to give a full, detailed, thorough-going, comprehensive and exhaustive description of the unique features and startling characteristics of all the mathematical lore in question.

This can and will be done in the subsequent volumes of this series (dealing seriatim and in extenso with all the various portions of all the various branches of mathematics).

14. We may, however, at this point, draw the earnest attention of everyone concerned to the following salient items thereof:

   i. The Sutras (aphorisms) apply to and cover each and every part of each and every chapter of each and every branch of mathematics (including arithmetic, algebra, geometry—plane and solid, trigonometry- plane and spherical, conics—geometrical and analytical, astronomy, calculus—differential and integral etc., etc. In fact, there is no part of mathematics, pure or applied, which is beyond their jurisdiction;

   ii. The Sutras are easy to understand, easy to apply and easy to remember; and the whole work can be truthfully summarised in one word "mental"!

   iii. Even as regards complex problems involving a good number of mathematical operations (consecutively or even simultaneously to be performed), the time taken by the Vedic method will be a third, a fourth, a tenth or even a much smaller fraction of time required according to the modern (i.e. current) Western methods;

   iv. And, in some very important and striking cases, sums requiring 30, 50, 100 or even more numerous and cumbrous "steps" of working (according to the current Western methods) can be answered in a single and simple step of work by the Vedic method! And little children (of only 10 or 12 years of age) merely look at the sums written on the blackboard (on the platform) and immediately shout out and dictate the answers from the body of the convocation hall (or other venue of demonstration). And this is because, as a matter of fact, each digit automatically yields its predecessor and its successor! and the children have merely to go on tossing off (or reeling off) the digits one after another (forwards or backwards) by mere mental arithmetic (without needing pen or pencil, paper or slate etc.)!

   v. On seeing this kind of work actually being performed by the little children, the doctors, professors and other "big-guns" of mathematics are wonder struck and exclaim:- "Is this mathematics or magic?" And we invariably answer and say: "It is both. It is magic until you understand it; and it is mathematics thereafter"; and then we proceed to substantiate and prove the correctness of this reply of ours!

   vi. And as regards the time required by the students for mastering the whole course of Vedic Mathematics as applied to all its branches, we need merely state from our actual experience that 8 months (or 12 months) at an average rate of 2 or 3 hours per day should suffice for completing the whole course of mathematical studies on these Vedic lines instead of 15 or 20 years required according to the existing systems of Indian and also of foreign universities.

15. In this connection, it is a gratifying fact that unlike some so-called Indologists (of the type hereinabove referred to) there have been some great modern mathematicians and historians of mathematics (like Prof. G.P. Halstead, Professor Ginsburg, Prof. De Morgan, Prof. Hutton etc.,) who have, as truth-seekers and truth-lovers, evinced a truly scientific attitude and frankly expressed their intense and whole-hearted appreciation of ancient India's grand and glorious contributions to the progress of mathematical knowledge (in the Western hemisphere and elsewhere).

16. The following few excerpts from the published writings of some universally acknowledged authorities in the domain of the history of mathematics, will speak eloquently for themselves:

   i. On page 20 of his book *On the Foundation and Technique of Arithmetic,* we find Prof. G.P. Halstead saying, "The importance of the creation of the zero mark can never be exaggerated. This giving of airy nothing not merely a local habitation and a name, a picture but helpful power is the characteristic of the Hindu race whence it sprang. It is like coining the Nirvana into dynamos. No single mathematical creation has been more potent for the general on-go of intelligence and power."

ii. In this connection, in his splendid treatise on "The present mode of expressing numbers" (*The Indian Historical Quarterly*, Vol. 3, pages 530-540) B.B. Dutta says, "The Hindus adopted the decimal scale very early. The numerical language of no other nation is so scientific and has attained as high a state of perfection as that of the ancient Hindus. In symbolism they succeeded with ten signs to express any number most elegantly and simply. It is this beauty of the Hindu numerical notation which attracted the attention of all the civilised peoples of the world and charmed them to adopt it".

iii. In this very context, Prof. Ginsburg says:—"The Hindu notation was carried to Arabia about 770 A.D. by a Hindu scholar named Kanka who was invited from Ujjain to the famous court of Baghdad by the Abbaside Khalif Al-Mansur. Kanka taught Hindu astronomy and mathematics to the Arabian scholars; and, with his help, they translated into Arabic the Brahma-Sphuta-Siddhanta of Brahma Gupta. The recent discovery by the French savant M.F. Nau proves that the Hindu numerals were well known and much appreciated in Syria about the middle of the 7th Century A.D. (Ginsburg's "New light on our numerals", *Bulletin of the American Mathematical Society*, Second Series, Vol. 25, pages 366-369).

iv. On this point, we find B.B. Dutta further saying: "From Arabia, the numerals slowly marched towards the West through Egypt and Northern Arabia; and they finally entered Europe in the 11th Century. The Europeans called them the Arabic notations, because they received them from the Arabs. But the Arabs themselves, the Eastern as well as the Western, have unanimously called them the Hindu figures. (Al-Arqan-Al-Hindu)."

17. The above-cited passages are, however, in connection with, and in appreciation of India's invention of the "Zero" mark and her contributions of the 7th century A.D. and later to world mathematical knowledge.

In the light, however, of the hereinabove given detailed description of the unique merits and characteristic excellences of the still earlier Vedic Sutras dealt with in the 16 volumes of this series, the conscientious (truth-loving and truth-telling) historians of mathematics (of the lofty eminence of Prof. De Morgan etc.) have not been guilty of even the least exaggeration in their candid admission that "even the highest and farthest reaches of modern Western mathematics have not yet brought the Western world even to the threshold of Ancient Indian Vedic Mathematics".

18. It is our earnest aim and aspiration, in these 16 volumes, to explain and expound the contents of the Vedic Mathematical Sutras and bring them within the easy intellectual reach of every seeker after mathematical knowledge.

CONTENTS

Chapter 1

INTRODUCTION TO THE VERTICALLY AND CROSSWISE SUTRA

We look in this chapter at a number of applications of this formula in multiplication, division, squaring, square roots etc. Most of the methods described will be used in subsequent chapters.

CROSS-PRODUCTS AND MULTIPLICATION

1) For the product of two 1-figure numbers
   we simply multiply the figures; this is a vertical product.

$$\begin{array}{r} 2 \\ 3 \ \times \\ \hline 6 \end{array}$$

2) For the product of two 2-figure numbers we
   a) take the vertical product on the right: 2.1 = **2**;
   b) take the cross-product: 4.1 + 2.3 = **10**, put down 0 and carry 1;
   c) take the vertical product on the left: 4.3 = **12**,
      then 12 plus the carried 1 = 13.

$$\begin{array}{r} 4 \quad 2 \\ 3 \quad 1 \ \times \\ \hline 1\,3\,{}_{1}0\,2 \end{array}$$

Thus **42 × 31 = 1302**.

We can see that this simple and symmetrical procedure works because the product of the units digits in the sum will give the units digit of the answer, the product of tens times units and units times tens in the sum (plus any carried figures) will give the tens digits of the answer and the product of the tens digits in the sum (plus any carried figures) will give the hundreds digits of the answer.

In carrying out this sum we obtained 3 **vertical products or cross-products**.

We can write:

$$CP\,{}^{4}_{3}\binom{2}{1} = 2, \quad CP\binom{4\ \ 2}{3\ \ 1} = 10, \quad CP\binom{4}{3}{}^{2}_{1} = 12,$$

Please note that in this chapter we sometimes use a dot in place of a multiplication sign.
E.g. $2.7 = 2 \times 7$.

3) Similarly for the product
$$\begin{array}{r} 3\ 2 \\ 4\ 7\ \times \\ \hline 1{,}5{,}0{,}4 \end{array}$$

the 3 cross-products are:

a) $CP{}^{3}_{4}\!\left(\!\begin{array}{c}2\\7\end{array}\!\right) = 2.7 =$ $\qquad$ 1 4 $\qquad$ i.e. $\vdots\ |$

b) $CP\!\left(\!\begin{array}{cc}3 & 2\\ 4 & 7\end{array}\!\right) = 3.7 + 2.4 =$ $\qquad$ 2 9 $\qquad$ $\times$

c) $CP\!\left(\!\begin{array}{c}3\\4\end{array}\!\!\begin{array}{c}2\\7\end{array}\!\right) = 3.4 =$ $\qquad$ $\underline{1\ 2\qquad}$ $\qquad$ $|\ \vdots$

Therefore $\underline{32 \times 47} =$ $\qquad$ 1 5 0 4

4) Now consider the product 302×514.
Here we have 5 cross-products:

a) $CP{}^{30}_{51}\!\left(\!\begin{array}{c}2\\4\end{array}\!\right) = 2.4 =$ $\qquad$ 8 $\qquad$ $\vdots\ \vdots\ |$

b) $CP{}^{3}_{5}\!\left(\!\begin{array}{cc}0 & 2\\ 1 & 4\end{array}\!\right) = 0.4 + 2.1 =$ $\qquad$ 2 $\qquad$ $\vdots\ \times$

c) $CP\!\left(\!\begin{array}{ccc}3 & 0 & 2\\ 5 & 1 & 4\end{array}\!\right) = 3.4 + 0.1 + 2.5 =$ $\qquad$ 2 2 $\qquad$ $\times\!\!\!\times$

d) $CP\!\left(\!\begin{array}{cc}3 & 0\\ 5 & 1\end{array}\!\!\begin{array}{c}2\\4\end{array}\!\right) = 3.1 + 0.5 =$ $\qquad$ 3 $\qquad$ $\times\ \vdots$

e) $CP\!\left(\!\begin{array}{c}3\\5\end{array}\!\!\begin{array}{c}02\\14\end{array}\!\right) = 3.5 =$ $\qquad$ $\underline{1\ 5\qquad}$ $\qquad$ $|\ \vdots\ \vdots$

Therefore $\underline{302 \times 514} =$ $\qquad$ 1 5 5 2 2 8

5) $3251 \times 7604 = \underline{24720604}$

$$\begin{array}{r} 3\quad 2\quad 5\quad 1 \\ 7\quad 6\quad 0\quad 4 \\ \hline 2\ 4\ {}_{3}7\ {}_{5}2\ {}_{5}0\ {}_{1}6\ {}_{2}0\ 4 \end{array} \qquad \text{or} \qquad \begin{array}{r} 3\ \bullet\ 2\ \bullet\ 5\ \bullet\ 1 \\ 7\quad\ \ 6\quad\ \ 0\quad\ \ 4 \\ \hline 2\ 4\ {}_{3}7\ {}_{5}2\ {}_{5}0\ {}_{1}6\ {}_{2}0\ 4 \end{array}$$

We can write the answer straight down.

An alternative spacing is shown on the right. This has the advantage of enhancing the symmetry because each answer digit is put exactly under the centre of its product-pattern. Thus for $5.4 + 1.0$ the centre is in the middle of these four digits and the answer goes below this point- as the sum proceeds these centre points move across the sum.

The cross-products, which can be evaluated mentally, are:

a) CP $\begin{matrix} 3\ 2\ 5 \\ 7\ 6\ 0 \end{matrix} \binom{1}{4} = 1.4 =$ 4   : : :]

b) CP $\begin{matrix} 3\ 2 \\ 7\ 6 \end{matrix} \begin{pmatrix} 5\ 1 \\ 0\ 4 \end{pmatrix} = 5.4 + 1.0 =$ 2 0   : : ✕

c) CP $\begin{matrix} 3 \\ 7 \end{matrix} \begin{pmatrix} 2\ 5\ 1 \\ 6\ 0\ 4 \end{pmatrix} = 2.4 + 5.0 + 1.6 =$ 1 4   : ✳

d) CP $\begin{pmatrix} 3\ 2\ 5\ 1 \\ 7\ 6\ 0\ 4 \end{pmatrix} = 3.4 + 2.0 + 5.6 + 1.7 =$ 4 9   ✳

e) CP $\begin{pmatrix} 3\ 2\ 5 \\ 7\ 6\ 0 \end{pmatrix} \begin{matrix} 1 \\ 4 \end{matrix} = 3.0 + 2.6 + 5.7 =$ 4 7   ✳ :

f) CP $\begin{pmatrix} 3\ 2 \\ 7\ 6 \end{pmatrix} \begin{matrix} 5\ 1 \\ 0\ 4 \end{matrix} = 3.6 + 2.7 =$ 3 2   ✕ : :

g) CP $\begin{pmatrix} 3 \\ 7 \end{pmatrix} \begin{matrix} 2\ 5\ 1 \\ 6\ 0\ 4 \end{matrix} = 3.7 =$ <u>2 1   </u>   ] : : :

                                   2 4 7 2 0 6 0 4

6) 12131 × 20412 = <u>247617972</u>

$$
\begin{array}{r}
1\ 2\ 1\ 3\ 1 \\
2\ 0\ 4\ 1\ 2 \\
\hline
2\ 4\ 7\ {}_16\ {}_11\ {}_17\ 9\ 7\ 2
\end{array}
\quad \text{or} \quad
\begin{array}{r}
1 \quad 2 \quad 1 \quad 3 \quad 1 \\
2 \quad 0 \quad 4 \quad 1 \quad 2 \\
\hline
2\ 4\ 7\ {}_16\ {}_11\ {}_17\ 9\ 7\ 2
\end{array}
$$

There are 9 cross-products.

7) 6341 × 32 = <u>202912</u>

If, as here, the numbers being multiplied do not have the same number of figures we can insert zeros:

$$
\begin{array}{r}
6\ 3\ 4\ 1 \\
0\ 0\ 3\ 2 \\
\hline
2\ 0\ {}_2 2\ {}_1 9\ {}_1 1\ 2
\end{array}
$$

This method of multiplication is very easy and with a little practice it can also be very quick.

Exercise A (answers will be found on page 17)

1. 23 × 34   2. 41 × 42   3. 34 × 43   4. 61 × 73   5. 234 × 432
6. 334 × 236   7. 712 × 403   8. 676 × 425   9. 1234 × 2012   10. 3451 × 2233
11. 3434 × 4321   12. 4234 × 24   13. 2372 × 23   14. 4545 × 123   15. 30123 × 52
16. 35703 × 321   17. 4554 × 7007   18. 928161 × 61   19. 999 × 888   20. 52025 × 12345

THE VINCULUM

Arithmetic operations can be greatly simplified by the use of the vinculum. In fact it is never really necessary in the Vedic system to use digits over 5 (i.e. 6, 7, 8, 9).

8) As 38 is close to 40 we may use $4\overline{2}$ (i.e. $40 - 2$) instead of 38.

Similarly:

9) $199 = 20\overline{1}$ 10) $2882 = 3\overline{12}2$ 11) $318297 = 32\overline{2}30\overline{3}$

12) $278987 = 32\overline{1013}$ 13) $2\overline{6}341 = 14341$ 14) $345\overline{6} = 3344$

15) $10\overline{22}\overline{3} = 9817$

The Vedic formula ALL FROM 9 AND THE LAST FROM 10 helps us to get or remove these vinculum figures: in Example 12 the numbers under the vinculums are obtained by taking each of the large digits 78987 from 9 and the last from 10.

Similarly in Example 14 we take 5 and 6 from 9 and 10 respectively to clear the vinculums.

In this way, not only are large figures avoided, but 0 and 1 appear twice as frequently as they otherwise would, and these are particularly easy digits to work with.
A third advantage is that numbers often partly or wholly cancel themselves, as will be seen below. We will be making extensive use of the vinculum.

16) $39 \times 49 = \underline{1911}$ $39 \times 49 = 4\overline{1} \times 5\overline{1}$:

$$
\begin{array}{rrrr}
 & & 4 & \overline{1} \\
 & & 5 & \overline{1} \\
\hline
2 & 0 & \overline{9} & 1 \\
= 1 & 9 & 1 & 1 \\
\end{array}
$$

17) $291 \times 388 = 3\overline{1}1 \times 4\overline{12} = \underline{112908}$:

$$
\begin{array}{rrrrrr}
 & & 3 & \overline{1} & 1 \\
 & & 4 & \overline{1} & \overline{2} \\
\hline
1 & 2 & \overline{7} & \overline{1} & 1 & \overline{2} \\
= 1 & 1 & 2 & 9 & 0 & 8 \\
\end{array}
$$

18) The vinculum is also very useful for division. For example $41 \div 6$ is close to 7 so we may say that $41 \div 6 = 7$ remainder $\overline{1}$.

Similarly:
19) $61 \div 9 = 7$ rem $\overline{2}$ 20) $13 \div 15 = 1$ rem $\overline{2}$ 21) $29 \div 6 = 5$ rem $\overline{1}$

Exercise B

1. 19×33	2. 39×41	3. 28×39	4. 191×31	5. 218×32
6. 27×49	7. 188×123	8. 129×189	9. 108×199	10. 298×689
11. $35 \div 6$	12. $52 \div 9$	13. $18 \div 20$	14. $90 \div 31$	15. $77 \div 78$

ALGEBRAIC PRODUCTS

22) $(3x + 4)(2x + 5)$

We can use the same VERTICAL AND CROSSWISE pattern here:

$$\begin{array}{r} 3x \ +4 \\ 2x \ + \ 5 \ \times \\ \hline 6x^2 \ + \ 23x \ + \ 20 \end{array}$$

where $20 = 4 \times 5,$
             $23x = 3x.5 + 4.2x,$
             $6x^2 = 3x.2x.$

23) $(2x^2 - 3x - 5)(x^2 - 1)$

$$\begin{array}{r} 2x^2 \ - \ 3x \ - \ 5 \\ x^2 \ + \ 0x \ - \ 1 \ \times \\ \hline 2x^4 - 3x^3 - 7x^2 + 3x \ + \ 5 \end{array}$$

This is similar to the product of two 3-figure numbers- cf. Example 4.

LEFT TO RIGHT CALCULATIONS

It is often important, and especially useful in mental mathematics, to obtain the figures of an answer from left to right, i.e. to get the most significant figure first, then the next most significant figure and so on. Normally only divisions are done in this way.

Addition

24)

$$\begin{array}{r} 3 \ 6 \ 7 \\ 9 \ 8 \ 5 \ + \\ \hline 1_2 \end{array}$$

In this addition sum, if we add the left-hand column we get 3+9=12. Since there will be a carried figure from the next column that will affect this total, we put down only the 1 and carry the 2 forwards, as shown. The middle column adds up to 14. To this we add the carried 2, **as 20**, to get 34 and put down 3_4 as shown.

$$\begin{array}{r} 3 \ 6 \ 7 \\ 9 \ 8 \ 5 \ + \\ \hline 1_2 3 \ _4 5 \ 2 \end{array}$$

Then adding the right-hand column we get 12, to which we add the carried 4, **as 40**, to get 52- which we put down.

25)   7 3 4 1
         2 3 8 7
         7 8 7 8 +
     1₆ 7₄ 5₉ 10 6
   = 1 7 6 0 6

We could use the vinculum in this sum,
and put $6\bar{1}$ instead of 59 in the third column:

     7 3 4 1
     2 3 8 7
     7 8 7 8 +
   1 ₆7 ₄6 ᵢ0 6

26)   3 9 6
         8 8 7
         7 7 8 +
     2 ₂0 ₄6 1

Subtraction

27)   7 ¹3 2
      1 8 4 —
      5

In finding 732 – 184 we subtract in the left-hand column:
7–1=6, but since a unit is needed for the next column
(3 is less than 8) we put down only 5 and place the other
unit as shown.

      7 ¹3 ¹2
      1 8 4 -
      5 4 8

Then 13–8=5, but again a unit is needed in the next column
(2 is less than 4), so we put down only 4.
Finally 12–4=8.

28)   5 ¹4 ¹7 2 ¹0 ¹3
      1 9 9 1 7 8 -
      3 4 8 0 2 5

Multiplication

29)   3457 × 8 = 27656

We begin on the left: 3.8=24, put as shown.
Then 4.8=32: add the carried 4, **as 40**, 32+40=72, as shown.
Then 5.8=40: add the carried 2, as 20, 40+20=60, as shown.
Finally 7.8=56, 56+0=56, which we put down.

         3 4 5 7
                  8 ×
     2 ₄7 ₂6 ₀5 6

30)   86379 × 6 = 518274

Here we make use of the vinculum.

$8.6 = 48 = 5\bar{2}$, put $5_{\bar{2}}$.

$6.6 = 36, 36 + \overline{20} = 16$, put 1_6.

$3.6 = 18 = 2\bar{2}, 2\bar{2} + 60 = 8\bar{2}$, put $8_{\bar{2}}$ etc.

         8 6 3 7 9
                     6 ×
     5 ₂1 ₆8 ₂2 ₂7 4

31)   1 6 8 8
$\underline{3}$ ×
0₃5$_{\bar{2}}$0₄6 4

32) 63 × 74 = <u>4662</u>

6 3
$\underline{7\ 4}$
4₂6₅6 2

6.7 = 42, put down 4 and carry 2.
6.4 + 3.7 = 45, add the carried 2, as 20:
45 + 20 = 65, put down 6 and carry 5.
3.4 = 12, add the carried 5, as 50:
12 + 50 = 62, put down 62.

33) 4 3 7
$\underline{5\ 2\ 6}$
2₀2₃9₅8₂6 2

34) 2 4
$\underline{3\ 2}$
0₆7₆6 8

2.3 = 6 (a single figure), carry 6.
2.2 + 4.3 = 16, 16 + 60 = 76.
4.2 = 8, 8 + 60 = 68.

Using the Vinculum

35) 8 2
$\underline{6\ 8}$
5$_{\bar{2}}$5₆7 6

Since 8.6 = 48 we would prefer not to add 80 at the next
step, so we can write 48 as $5\bar{2}$, put down 5 and carry $\bar{2}$

36) 39 × 69 = <u>2691</u>

4 $\bar{1}$
$\underline{7\ \ \bar{1}}$
2₈6₉9 1

Here we may change the figures in the sum to remove the
large digits.

37) Find 345243 × 761283 to 3 significant figures.

3 4 5 2 4 3
$\underline{7\ 6\ 1\ 2\ 8\ 3}$
2₁6$_{\bar{4}}$2₂7₄···

∴ 345243 × 761283 = 2.63 × 10¹¹ to 3 S.F.

Exercise C

1. 4376 × 6	2. 7676 × 7	3. 8243 × 9	4. 43 × 63	5. 48 × 61
6. 26 × 68	7. 91 × 53	8. 821 × 321	9. 97 × 48	10. 634 × 183

MULTIPLYING WITH GROUPS OF FIGURES

38)   123 × 112   Since the numbers begin with 12 and 11 we can consider each of these as a group. Then from right to left:

$$\begin{array}{r} (12)\ 3 \\ \underline{(11)\ 2} \\ 1\,3\,7{,}7\,6 \end{array}$$

3.2 = 6, put down 6.
12.2 + 3.11 = 57, put down 7 and carry 5.
12.11 = 132, 132 + 5 = 137.

39)   312 × 411   If the pairs are on the right of the numbers we put down the answer digits in groups of two:

$$\begin{array}{r} 3\ (12) \\ \underline{4\ (11)} \\ 1\,2\,/\,8\,2\,/{,}3\,2 \end{array}$$

12.11 = 132, put down 32 and carry 1.
3.11 + 12.4 = 81, 81 + 1 = 82, put down 82.
3.4 = 12, put down 12.

40)   1112 × 1212

$$\begin{array}{r} (11)\ (12) \\ \underline{(12)\ (12)} \\ 1\,3\,4\,_{,}7\,7\,_{,}4\,4 \end{array}$$

This method can be extended to products like:

$$\begin{array}{ccc} (21) & (32) & (43) \\ \underline{(11)} & \underline{(23)} & \underline{(34)} \\ & & \end{array} \qquad \begin{array}{cc} (213) & (123) \\ \underline{(312)} & \underline{(112)} \\ & \end{array}$$

and so on.

PRODUCTS OF THREE NUMBERS

41)   21 × 43 × 65 = <u>58695</u>

$$\begin{array}{r} 2\ 1 \\ 4\ 3 \\ \underline{6\ 5} \\ 5\ 8\,_{10}6\,_{6}9\,_{1}5 \end{array}$$

For the units digit: 1.3.5 = 15.
For the tens digit: 2.3.5 + 4.1.5 + 6.3.1 = 68.
For the hundreds: 1.4.6 + 3.2.6 + 5.4.2 = 100.
For the thousands: 2.4.6 = 48.

MOVING MULTIPLIER

Multiplying a long number by a short number.

42) 61261 × 43

$$6 \quad 1 \quad 2 \quad 6 \quad 1$$
$$\underline{\qquad\qquad 4 \quad 3}$$
$$2 \quad 6_{\,2}3_{\,1}4_{\,3}2_{\,2}2 \quad 3$$

We do not have to insert zeros when one number is shorter than the other: we can move the multiplier along its row, or imagine it to be moved, multiplying crosswise as we go.
First, in the position shown, the vertical product on the right is 1.3 = 3. The cross-product gives 22 (put down 2 carry 2).

Then we move the 43 one place to the left so that it is under 26 and cross-multiply to get 30, then 30 + 2 = 32.
We continue this process, moving the multiplier one place to the left each time. In the final position we also take the vertical product on the left.

43) Find 8273641 × 213 to 4 S.F.

$$8 \quad 2 \quad 7 \quad 3 \quad 6 \quad 4 \quad 1$$
$$\underline{2 \quad 1 \quad 3}$$
$$1_{\,6}7_{\,2}6_{\,0}2 \quad 2 \,_{\bar 1}3_{\,\bar 4}\ldots$$

Here we calculate from left to right; two shifts are needed.

$\therefore\ 1.762 \times 10^9$ to 4 S.F.

Exercise D

1. 114 × 121	2. 122 × 123	3. 311 × 411	4. 603 × 412	5. 1213 × 1112
6. 1321 × 23	7. 3123 × 33	8. 21012 × 41	9. 357 × 62	10. 121314 × 31
11. 6237 × 123	12. 30405 × 321	13. 8213 × 263 to 3 S.F.	14. 36214 × 81 to 3 S.F.	15. 67176 × 83262 to 3 S.F.

NUMBER OF ZEROS AFTER THE DECIMAL POINT

44) 0.008 × 0.009 = 0.000072

45) 0.0812 × 0.032

$$0.0 \, 8 \, 1 \, 2$$
$$\underline{\qquad 0.0 \, 3 \, 2}$$
$$0.0 \, 0 \, 2 \, 5 \, 9 \, 8 \, 4$$

We note from these two examples that the number of zeros after the decimal point in the answer is the same as the number of zeros after the decimal points in the numbers being multiplied.

46) 0.004 × 0.02 = 0.00008

But here, since 4.2 = 8, i.e. a single digit, there is one more zero in the answer.

This is useful when calculating products from left to right.

ARGUMENTAL DIVISION

47) $3468 \div 72 = \underline{48 \, R12}$

                    A B
                    7 2
              3 4$_6$6$_2$8

We know that the vertical product A×7 on the left must account for most of the 34 on the left of the product 3468, and therefore A = 4. This accounts for 2800 of the 3400 so that there is a remainder of 6 (hundreds) as shown.

We now have effectively got 66 in the tens column.

In the cross-product A×2 + B×7, which must account for most of this 66, we know A×2 = 8 so that B×7 = 58 and B is therefore 8 with a remainder of 2.

We now have 28 effectively in the units column and since the vertical product B×2 = 16 the remainder is 12.

48)   A B C
                  7 1 2
            2 2 $_1$8 $_1$5 5 8

We see that $228558 \div 712 = 321 \, R6$.

49)   A B C D
                        9 1
            2 1 · 3 4 5 6

Here $213456 \div 91 = 2345 \, R61$.

We see from these examples that this division is just the reverse of the simple Vedic process of multiplication.

STRAIGHT DIVISION

These divisions can be performed with much greater ease with the following arrangement.

50) $3468 \div 72 = 48 \, R12$

              2) 3 4 6⁄ 8⁄
         7   6⁄ 2⁄
              ———————————————
                    4⁄ 8⁄ 12

We set the sum out as shown. Then 34÷7 = 4 R6, as shown. Then 4 (in the answer)×2 (raised ON THE FLAG) = 8, 66–8 = 58 and 58÷7 = 8 R2, placed as shown.

Next 8 (in the answer)×2 (on the flag) = 16, 28–16 = 12 placed as shown.

It will be seen that these operations are identical to those in Example 47, but the procedure is simpler since at each stage we just multiply the last answer digit by the flag number, subtract the product from the next dividend, divide by the (single digit) divisor and put down the result.

The sum thus proceeds in distinct stages indicated by the diagonal lines above.

51)

$$8 \quad 1\,)\,2\;1\;3\;\,1\;\,4\;\,1$$
$$\quad \quad 5\;3\;1\;3$$
$$\quad \overline{\;2\;6\;3\;1\;30}$$

As in the previous example, having one figure on the flag we mark off one figure on the right of the dividend to indicate the remainder column.

52) Using the vinculum:

$$6 \quad 3\,)\,3\;7\;3\;7\;|\;3$$
$$\quad \;\;1\;\;\bar5\;\;|\;\bar1$$
$$\quad \overline{\;\;6\;0\;\;\bar7\;|\;14}$$

$37 \div 6 = 6\ R1$.

$6.3 = 18,\ 13 - 18 = \bar5,\ \bar5 \div 6 = 0\ R\bar5$.

Our next dividend is $\bar57$ or $\overline{43}$.

$0.3 = 0,\ \overline{43} - 0 = \overline{43},\ \overline{43} \div 6 = \bar7\ R\bar1$ as shown.

$\bar7.3 = \overline{21},\ \bar1 3 - \overline{21} = \bar7 + 21 = 14$.

53) Avoiding the vinculum:

$$6 \quad 3\,)\,3\;7\;3\;7\;|\;3$$
$$\quad \;\;7\;\;4\;|\;2$$
$$\quad \overline{\;\;5\;9\;3\;|\;14}$$

If we wish to avoid the vinculum then instead of $37 \div 6 = 6\ R1$, we may say $37 \div 6 = 5\ R7$ and continue as usual.

54) Vinculum on the flag:

If when dividing by 49 we put 9 on the flag the subtractions are likely to be large.

$$5 \quad \bar1\,)\,6\;2\;3\;1\;|\;2$$
$$\quad \;1\;3\;0\;|\;3$$
$$\quad \overline{\,1\;2\;7\;1\;|\;33}$$

If we use $5\bar1$ as a divisor instead of 49 we **add** the product of the last answer digit and the flag number at each step instead of subtracting.

Thus $6 \div 5 = 1\ R1,\ \bar1.1 = \bar1$,

$12 - \bar1 = 12 + 1 = 13,\ 13 \div 5 = 2\ R3$ etc.

55) $54545 \div 29$

$$3 \quad \bar1\,)\,5\;4\;5\;4\;|\;5$$
$$\quad \;2\;1\;2\;|\;1$$
$$\quad \overline{\,1\;8\;7\;10\,|\,25}$$
$$\quad =\;1\;8\;8\;\;0\;R25$$

Here we get 10 in the last place of the answer, so we carry 1 over to the left (but multiply the flag digit by **10** at the last step).

56) Decimalising the remainder: $333 \div 73$

$$7 \quad 3\,)\,3\;3\;3\,.\,0\;0\;0$$
$$\quad \;5\;6\;3\;5$$
$$\quad \overline{\,4\,.\,5\;6\;1\ldots}$$

The decimal point occupies the same position as the vertical line would have, and we just keep dividing in the usual way.

57)   2-digit divisor: 776655÷203

```
      3)7 7 6 6 5 5.0
   20        17 7  12  19 0
          ─────────────────
          3 8 2 5.9...
```
We can use a 2-digit divisor, 20 in this case, if we wish.

58)   Two figures on the flag: 2999222÷713

```
     13)2 9 9 9 2 2 2
   7          1  1.  5  4  36
          ─────────────────
          4 2 0 6 344
```
$29 \div 7 = 4$ R1.
Then 4.1 (on the flag) = 4, 19–4 = 15, $15 \div 7 = 2$ R1, as shown.
We then cross-multiply the two answer digits by the two flag digits: 1.2 + 3.4 = 14, 19–14 = 5, $5 \div 7 = 0$ R5, as shown.

Again multiply the last two answer digits by the flag digits: 1.0 + 3.2 = 6, 52–6 = 46, $46 \div 7 = 6$ R4, as shown.
Cross-multiply again: 1.6 + 3.0 = 6, 42 – 6 = 36, put as shown.
Then finally take the last (vertical) product of 6 (in the answer)×3 (on the flag) = 18, 362–18 = 344, the remainder.

59)   Three figures on the flag: 987987 ÷ 8123

```
    123)9 8 7 9 8 7
   8        1  1  5 51 510
          ─────────────────
          1 2 1 5104
```
$9 \div 8 = 1$ R1, put it down.
1.1 (on the flag) = 1, 18–1 = 17, $17 \div 8 = 2$ R1, put it down.
1.2 + 2.1 = 4, 17–4 = 13, $13 \div 8 = 1$ R5, put it down.
Next we cross-multiply the three answer figures with the three flag digits: 1.1 + 2.2 + 3.1 = 8, 59–8 = 51, put it down.
2.1 + 3.2 = 8, 518 – 8 = 510, put it down.
Finally 3.1 = 3, 51070 – 3 = 5104.

60)   Decimalised remainder: 54341 ÷ 7103

```
    103)5 4 3 4 1.0 0 0 0 0
   7         5 .4 3 5. 4 0 3̄ 0
          ──────────────────────
          7.6 5 0 4 3 0 6̄..
```

61)   Using the vinculum: 34567 ÷ 6918

```
    1̄2̄2)3 4 5 6 7.0 0 0
   7         6 6 4. 4 4 3
          ──────────────────
          4.9 9 6 6 7...
```

62) Find 877778 ÷ 819976 to 5 decimal places

$$2002\bar{4})8\ 7\ 7\ 7\ 7\ 8\ .\ 0$$
$$8 \qquad \underline{0\ \ 5\ \ 1\ \ 3\ \ \bar{1}\ \ 0}$$
$$\overline{1\ .\ 0\ \ 7\ \ 0\ \ 5\ \ \bar{1}}$$

Exercise E

1. 92)2222   2. 73)40342   3. 81)30405   4. 53)33221   5. 112)55443
6. 914)760559   7. 923)62054   8. 9123)76.543   9. 191)19843   10. 88)9.8765

The same division process can be applied to algebraic division (see Chapter 13).

RECIPROCALS

We can also use argumental division to find reciprocals.

63) Find $\dfrac{1}{423}$

We are looking for a number which, when multiplied by 423 gives 1:

$$4\ \ 2\ \ 3$$
$$\underline{2\ .\ \ .\ \ .}$$
$$1\ .\ 0\ \ 0\ \ 0\ \ .\ \ .\ \ .$$

The first figure is clearly a 2, as shown above.
The cross-product in the first two columns now needs to be 20, because the first vertical product is 4×2 = 8, which is 2 short of the required 10.

So we now have:
$$4\ \ 2\ \ 3$$
$$\underline{2\ \ 4\ .\ \ .\ \ .}$$
$$1\ .\ 0\ \ 0\ \ 0\ \ .\ \ .\ \ .$$

The third figure of the answer now needs to be zero, as there were no carries from the previous step. The two products 2×4 and 3×2 give 14 and so the next figure will be negative: $\bar{3}$ or $\bar{4}$.
Choosing the last of these, and continuing for one more step gives:

$$4\ \ 2\ \ 3$$
$$\underline{2\ \ 4\ \ \bar{4}\ \ 4\ .\ \ .\ \ .}$$
$$1\ .\ 0\ \ 0\ \ 0\ \ .\ \ .\ \ .$$

So $\dfrac{1}{423} = 0.002364$ to 4 S.F.

64)   Find $\dfrac{2}{0.0442}$

Similarly here we find $\dfrac{1}{0.0221}$:

$$
\begin{array}{ccccccc}
2 & 2 & 1 & 0 & 0 & 0 \\
& 4 & 5 & 2 & 5 & . & . & . \\
\hline
1 & 0 & 0 & 0 & 0 & . & . & . \\
\end{array}
$$

So $\dfrac{2}{0.0442} = 45.25$ to 4 S.F.

> **SQUARING**

We make extensive use of the "Duplex", D, for squares and square roots.
The term has the following meaning.

For a 1-digit number D is its square,
for a 2-digit number D is twice their product,
for a 3-digit number D is twice the product of the outer pair + the square of the middle digit,
for a 4-digit number D is twice the product of the outer pair + twice the product of the inner
pair,

and so on.

Thus $D(3) = 9$, $D(42) = 16$, $D(127) = 18$, $D(2134) = 22$, $D(21312) = 19$.

65)   For 43^2 we find the duplexes of the first figure, the pair of figures and the last figure:
       $D(4) = 16$, $D(43) = 24$, $D(3) = 9$.
       $\therefore 43^2 = 16/_24/9 = \underline{1849}$

66)   For 67^2 $D(6) = 36$, $D(67) = 84$, $D(7) = 49$:
       $\therefore 67^2 = 36/_84/_49 = \underline{4489}$

67) 341^2 $D(3) = 9$, $D(34) = 24$, $D(341) = 22$, $D(41) = 8$, $D(1) = 1$:
       $\therefore 341^2 = 9/_24/_22/8/1 = \underline{116281}$

68) 1435^2 We find the duplexes of 1, 14, 143, 1435, 435, 35 and 5.
       $\therefore 1435^2 = 1/8/_22/_34/_49/_30/_25 = \underline{2059225}$

69) 21034^2 We find duplexes for 2, 21, 210, 2103, 21034, 1034, 034, 34, 4.
                    The duplex of 21034 is $2.(2.4) + 2.(1.3) + 0^2$.
                    $\therefore 21034^2 = 4/4/1/_12/_22/8/9/_24/_16 = \underline{442429156}$

If we do the calculation from right to left we can take up the carry figures as we go.

70) 542^2 Or, in a similar way to Examples 29–34 we can square from left to right taking
                    up carry figures (to which we add a zero) as we go.
                    $542^2 = 2_5\ 9_0\ 3_6\ 7_6\ 6\ 4 = \underline{293764}$

71) $(x + 3)^2 = \underline{x^2 + 6x + 9}$ We take duplexes of x, x+3 and 3.
                                        $D(x) = x^2$, $D(x+3) = 6x$ and $D(3) = 9$.

72) $(2x - 3y)^2 = \underline{4x^2 - 12xy + 9y^2}$

73) $(x + 2y + 3)^2 = \underline{x^2 + 4xy + 6x + 4y^2 + 12y + 9}$

| SQUARE ROOTS |

This is just the reverse of the squaring process.

74) $\sqrt{38837824} = \underline{6232}$ Here we observe 4 pairs of digits and therefore expect a
                                        4-figure answer if the number is a perfect square. Since the
       $\sqrt{38'_2\,8_4 3'_3\,7_8 8'_1\,24}$ first pair is 38 the first answer digit is 6, i.e. a=6, and the
                                        remainder is $38 - 6^2 = 2$, placed as shown.

       $= a\ b\ c\ d$ The next dividend is then 28. The duplex 2ab = 12b must
                                        account for most of this 28 so that b=2. The remainder, 4, is
       $= 6\ 2\ 3\ 2$ prefixed to the next figure as shown.

                                        The next dividend is 43, and the next duplex is $(2ac + b^2) =$
                                        12c + 4. We therefore subtract the 4: 43–4 = 39, 39÷12 = 3 R3,
                                        so that c=3.

                                        The next dividend is 37 and the duplex is $(2ad + 2bc) =$
                                        12d + 12, so 37-12 = 25, 25÷12 = 2 r1, therefore d=2.

                                        The next duplex is $(2bd + c^2) = 17$, 18–17 = 1; the next is
                                        2cd = 12, 12-12 = 0; and the last is $2^2 = 4$, 4–4 = 0, so that these
                                        last three duplexes account exactly for the remaining figures.

75) $\sqrt{28526281} = \underline{5341}$ This is similar to the example above and we may notice in both examples that at each step after the first we divide by **twice** the the first answer digit.

This allows the process to be simplified by setting up a chart like that shown opposite and dividing at each step by twice the first answer digit, after subtracting the duplex obtained from the figures which appear **after** the first figure.

$$\begin{array}{l} 2\ 8\ 5\ 2\ 6\ 2\ 8\ 1 \\ \ \ \ \ \ ^{3\ 5\ 3\ 2\ 0\ 0} \\ \hline 5/3\ 4\ 1.0\ 0\ 0 \end{array}$$

10)

i.e. put down 5 R3, then 35÷10 = 3 R5, put them down.
D(3) = 9, 52–9 = 43, 43÷10 = 4 R3, put them down.
D(34) = 24, 36–24 = 12, 12÷10 = 1 R2, put them down.

It will be seen that the steps are exactly as in the first method.
If we did not know the number was a perfect square and continued the process we would have:
D(341) = 22, 22–22 = 0, 0÷10 = 0 R0, put them down
D(3410) = 8, 8–8 = 0, 0÷10 = 0 R0, put them down.
D(34100) = 1, 1-1 = 0. And all successive duplexes are zero.

76) $\sqrt{1793} = 42.344\bar{1}\ldots$

$$\begin{array}{l} 1\ 7\ 9\ 3.0\ 0\ 0\ 0 \\ \ \ \ \ ^{1\ 3\ 5\ 6\ 3\ \bar{2}} \\ \hline 4/2\ .3\ 4\ 4\ \bar{1}\ldots \end{array}$$

8)

This is not an exact square and so the answer does not terminate.

77) Using the vinculum: $\sqrt{62} = 7.87401$

$$\begin{array}{l} 6\ 2.0\ 0\ 0\ 0\ 0\ 0 \\ \ \ \ ^{\bar{2}\ \bar{4}\ 7\ 0\ \bar{1}\ \bar{2}} \\ \hline 8.\bar{1}\ \bar{3}\ 4\ 0\ 1\ldots \end{array}$$

16)

Here, as 62 is close to the perfect square 64, we may take the first figure as 8 and have a remainder of $\bar{2}$.

Exercise F

Square the following:

1. 43	2. 73	3. 234	4. 712	5. 382
6. 4172	7. 6047	8. 63213	9. 34356	10. 491928

Find the square root:

11. 283024	12. 4489	13. 670761	14. 60516	15. 528529
16. 114244	17. 10784656	18. 55457809	19. 7890481	20. 3936256

Find the square root to 3 D.P.:

21. 2828	22. 3333	23. 2122	24. 234	25. 80908
26. 8	27. 10	28. 6.6	29. 48.81	30. 1.884

ANSWERS to exercises in Chapter 1.

Exercise A, page 3

1. 782	2. 1722	3. 1462	4. 4453	5. 101088
6. 78824	7. 286936	8. 287300	9. 2482808	10. 7706083
11. 14838314	12. 101616	13. 54556	14. 559035	15. 1566396
16. 11460663	17. 31909878	18. 56617821	19. 887112	20. 642248625

Exercise B, page 5

1. 627	2. 1599	3. 1092	4. 5921	5. 6976
6. 1323	7. 23124	8. 24381	9. 21492	10. 205322
11. 6 R$\bar{1}$	12. 6 R$\bar{2}$	13. 1 R$\bar{2}$	14. 3 R$\bar{3}$	15. 1 R$\bar{1}$

Exercise C, page 8

1. 26256	2. 53732	3. 74187	4. 2709	5. 2829
6. 1768	7. 4823	8. 263541	9. 4656	10. 116022

Exercise D, page 9

1. 13794	2. 15006	3. 127821	4. 248436	5. 1348856
6. 30383	7. 103059	8. 861492	9. 22134	10. 3760734
11. 767151	12. 9760005	13. 2160000	14. 2930000	15. 5590000000

Exercise E, page 13

1. 24 R14	2. 552 R46	3. 375 R30	4. 626 R43	5. 495 R3
6. 832.1214	7. 67.23077	8. 0.008390	9. 103.890	10. 0.112233

Exercise F, page 16

1. 1849	2. 5329	3. 54756	4. 506944	5. 145924
6. 17405584	7. 36566209	8. 3995883369	9. 1180334736	10. 241993157184
11. 532	12. 67	13. 819	14. 246	15. 727
16. 338	17. 3284	18. 7447	19. 2809	20. 1984
21. 53.179	22. 57.732	23. 46.065	24. 15.297	25. 284.443
26. 2.828	27. 3.162	28. 2.569	29. 6.986	30. 1.373

Chapter 2

COMBINED OPERATIONS OF ELEMENTARY ARITHMETIC

When there are two operations to be performed, such as multiplication and addition, the normal western procedure is first to carry out one of them, and then the other. With the Vedic system there is much more flexibility of sequence, and this often allows us to perform in one line operations of arithmetic or algebra which would otherwise take two or more lines. Indeed, the one-line mental calculation is a Vedic ideal; and the following methods show its practicality in this context.

But there is a further consideration. The reader may well be curious to see the various contributions different sutras can make to the same problem. We can, for instance, as will be shown in section (a) below, obtain sums of products in various ways. Four are shown here using the sutras BY ALTERNATE ELIMINATION AND RETENTION, VERTICALLY AND CROSSWISE, BY ADDITION AND BY SUBTRACTION and by using the sub-sutra PROPORTIONATELY. Which method or methods we choose depends on the context.

Practical applications of calculating sums of products are shown elsewhere in this book- as also of sums of squares, a method for which is given in section (b) below. Section (c) deals with combined addition and division.

(a) SUMS OF PRODUCTS

First method | Here we employ the sutra BY ALTERNATE ELIMINATION AND RETENTION.

Example 1   Calculate $6×9 + 7×8 + 4×3$

Each of the three products is a 2-digit number;
eliminating the first figure in each case we have $4 + 6 + 2$ (the last figures of 54, 56 and 12). Hence the last digit is 2, with 1 to carry.
Now we eliminate the last digits giving: 1 (carried) $+ 5 + 5 + 1 = 12$.
<u>Answer: 122</u>

Example 2   Calculate $7 \times 9 + 6 \times 4 + 5 \times 8$

Last digits give: $3 + 4 + 0 = 7$.
First digits give: $6 + 2 + 4 = 12$.
Answer: 127

Example 3 $6 \times 7 + 18 \times 9 + 14 \times 6 + 3 \times 8$

Here we observe that the four products in question are:
42, $2 \times 9^2 = 2 \times 81 = \mathbf{162}$, $2 \times 7 \times 6 = 2 \times 42 = \mathbf{84}$ and **24**.
Last digits give: $2 + 2 + 4 + 4 = 2$ and carry 1.
First digits give: 1 (carried) $+ 4 + 16 + 8 + 2 = 31$
Answer: 312

Note that for each product the last digit is given by the product of the last digits. But here in the last example we are beginning to stray into the territory of the second method, which will now be considered.

| **Second method** | This draws on the VERTICALLY AND CROSSWISE sutra. |

Example 4   Calculate $43 \times 72 + 54 \times 31$

We are going to carry out two "vertically and crosswise" multiplications simultaneously.
In the units place we have: $3 \times 2 + 4 \times 1 = 10$.
Write down 0 and carry 1.
So far we have:

$$
\begin{array}{ccc}
43 & & 54 \\
\times & + & \times \\
72 & & 31 \\
\hline
& & {}_1 0
\end{array}
$$

Next, in the tens place we have: $1 + 4 \times 2 + 3 \times 7 + 5 \times 1 + 4 \times 3 = 47$.
Note the scope here for employing the first method.
We now write down 7 and carry 4.
Thus, so far we have written:

$$
\begin{array}{ccc}
43 & & 54 \\
\times & + & \times \\
72 & & 31 \\
\hline
& {}_4 7 & {}_1 0
\end{array}
$$

Next, for the hundreds place, we have: 4 (carried) $+ 4 \times 7 + 5 \times 3 = 47$.
Hence the answer is 4770

A more convenient layout for the calculation is shown in the following example.

Example 5   Evaluate 213×426
$+ 354 \times 631$
$+ 652 \times 172$

$$
\begin{array}{ccc}
2 & 1 & 3 \\
4 & 2 & 6 \times \\
+ & & \\
3 & 5 & 4 \\
6 & 3 & 1 \times \\
+ & & \\
6 & 5 & 2 \\
1 & 7 & 2 \times \\
\hline
4\ 2\ _{10}6\ _{12}2\ _5 5\ _2 6
\end{array}
$$

The larger digits, such as the 7's, and 8's and 9's, can be avoided by using the vinculum as shown in Chapter 1.

Example 6   Evaluate $487 \times 519 + 714 \times 392$

Solution $\begin{array}{ccc} 5 & \bar{1} & \bar{3} \\ 5 & 2 & \bar{1} \\ + & & \\ 7 & 1 & 4 \\ 4 & \bar{1} & 2 \end{array}$

Answer $\underline{5\ 3\ 2\ 6\ 4\ 1}$

| Third method |

This time the sutra is BY ADDITION AND BY SUBTRACTION.

Example 7   Evaluate 6×9
$+ 7 \times 8$

Here we subtract 6 from 7 yielding 1, and add 8 to 9 to obtain 17.
This gives: 6×17
$+ 1 \times 8$

And now, $6 \times 7 + 8 = 42 + 8 = 50$ (0 carry 5), and $6 \times 1 + 5$ (carried) = **11**, whence the answer: $\underline{110}$

A more practical way of writing this is: $\begin{array}{c} ^{17} \\ 6 \times 9 \\ + 7 \times 8 = \underline{110} \\ _1 \end{array}$

Example 8   Evaluate $17 \times 13 + 14 \times 19 + 15 \times 17$

The complete working and solution can be written down as follows:

Step 1: $\qquad 17 \times \overset{28}{\cancel{13}}$ $\qquad$ Step 2: $\qquad 17 \times \overset{28}{\cancel{13}}{}^{0}$

$\qquad\qquad + 14 \times 19$ $\qquad\qquad\qquad + 14 \times \cancel{19}{}^{53}$

$\qquad\qquad + 15 \times \cancel{17}_0$ $\qquad\qquad\qquad + 15 \times \cancel{17}_0 = \underline{742}$

The steps are as follows:

1) 15 is added to 13 and the upper 17 subtracted from the lower 17.
   For brevity we can record this as: $13 + 15$ and $17 - 17$.
2) $28 - 2\times14$ and $19 + 2\times17$
3) $14 \times 53 = 742$ (vertically and crosswise).

Note that successive cancellations are allowed to wind around the original figure.
This method is well known in the context of second order determinants but is generally neglected in other contexts.
It can be useful when the multiples include complements, as in the next example.

Example 9 Evaluate $\qquad 43 \times 75$

$\qquad\qquad\qquad + 63 \times 49$

$\qquad\qquad\qquad + 27 \times 82$

Noting that 7 and 3 are complements we have:

$E = \qquad \overset{70}{\cancel{43}} \times 75$

$\qquad\quad + 63 \times 49$

$\qquad\quad + 27 \times \cancel{82}\ ^{7}$

This first step can be noted as: $43 + 27 = 80$ and $82 - 75 = 7$.
And now, noting that $27\times7 = 3\times9\times7 = 3\times63$, we have:

$E = \qquad 70 \times 75 \qquad = \qquad 70 \times 75 \qquad = \qquad 7 \times 750$

$\qquad\quad + 63 \times 49 \qquad\qquad\quad + 63 \times 52 \qquad\qquad + 7 \times 468$

$\qquad\quad + \underset{3}{\cancel{27}} \times \underset{63}{\cancel{7}}$

$\qquad\qquad\qquad\qquad\qquad\qquad\qquad\qquad\qquad = 7 \times 1218 = \underline{8526}$

Complementarity can be useful even if it does not include the final digit, as the next example demonstrates.

Example 10 Evaluate $\quad 6723 \times 5413 + 2281 \times 7632$

The **middle digits**, 72 and 28, are complementary, suggesting an addition.
The complete calculation can be written down as follows:

$\qquad \overset{9004}{\cancel{6723}} \times 5413$

$\qquad + 2281 \times \cancel{7632}\ ^{2221} \qquad = \qquad \underline{53,800,191}$

<u>Explanation</u>   Step 1: 6723 + 2281 and 7632 – 5413
Step 2: Multiply out and add, by "vertically and crosswise"

Note that the individual digits of 5413 and 7632 being fairly close to one another was also useful, giving a difference consisting of small digits. e.g. the initial figures, 5 and 7, are close, and so are the next ones, 4 and 6, etc.

In the next example the presence of like digits at the end of two different products suggests a subtraction.

Example 11   Evaluate $39 \times 537 + 29 \times 214$

<u>Solution</u> $\begin{array}{ccccc} 39 \times 537 & & 10 \times 537 & & 10 \times 2790 \\ + & = + & & = + \\ 29 \times 214 & & 29 \times 751 & & -1 \times 751 \end{array}$

$$= \underline{27,149}$$

The steps are: (1) $39 - 29$ and $214 + 537$
(2) $29 - 3 \times 10$ and $537 + 3 \times 751$
(3)   A subtraction (from 27900) completes the calculation.

An example of this type crops up in Chapter 7, when evaluating a logarithm.

| **Fourth method** | This provides an application of the sub-sutra PROPORTIONATELY.

Example 12   Evaluate $\begin{array}{r} 3 \times 2134 \\ + \ 4 \times 3261 \end{array}$
<u>Answer</u> $19{,}4{}_34{,}6$

Step 1: For units we have $3 \times 4 + 4 \times 1 = 16$; write down 6 and carry 1.
Step 2: In the tens place   1 (carried) $+ 3 \times 3 + 4 \times 6 = 34$;
write down 4 and carry 3, etc.

| **(b) SUMS OF SQUARES** |

The following method draws on the procedure for squaring given in Chapter 1 which is a special case arising from the VERTICALLY AND CROSSWISE Sutra.

Example 13   Evaluate $231^2 + 432^2 + 588^2$

The complete working and solution can be written as follows.

$$\begin{array}{ccc} 2 & 3 & 1 \\ 4 & 3 & 2 \\ 6 & \bar{1} & \bar{2} \end{array}$$

Answer   5 8 $_2$5 $_1$7 $_2$9

The steps are as follows, working from the right:

Step 1: For the units place we have $1^2 + 2^2 + \bar{2}^2 = 9$

Step 2: For the tens place $(3 \times 1 + 3 \times 2 + \bar{1} \times \bar{2}) \times 2 = 22$; write down 2
and carry 1.

Step 3: For the hundreds place
$2 \text{ (carried)} + 2 \times .2 \times 1 + 3^2 + 2 \times 4 \times 2 + 3^2 + 2 \times 6 \times \bar{2} + \bar{1}^2 = 17$;
write down 7 and carry 1.

Step 4: For the thousands place $1 \text{ (carried)} + (2 \times 3 + 4 \times 3 + 6 \times \bar{1}) \times 2 = 25$;
write down 5 and carry 2, and so on.

Working the same example from left to right the working and solution (with carried numbers written down) appear as follows.

$$\begin{array}{ccc} 2 & 3 & 1 \\ 4 & 3 & 2 \\ 6 & \bar{1} & \bar{2} \end{array}$$

Answer   5 $_6$8 $_4$5 $_5$7 $_2$9

The steps are as follows.

Step 1: For the 10,000's place we have $2^2 + 4^2 + 6^2 = 56$
We now have a choice between writing down the first digit on the left, 5, and carrying the 6 to the right, or writing down 6 and carrying $\bar{4}$.
If we recognise that the contribution to be added to this carried digit is small it pays to write down 5 and carry 6, as was done here.
If not, no great harm done—the answer simply comes out with a mixture of positive and negative digits.

$$\begin{array}{ccc} 2 & 3 & 1 \\ 4 & 3 & 2 \\ 6 & \bar{1} & \bar{2} \end{array}$$

5 $_6$

Step 2: For the 1000's place $(2 \times 3 + 4 \times 3 + 6 \times \bar{1}) \times 2 = 24$
This is equivalent to adding the duplexes $D(23) + D(43) + D(6\bar{1})$ as discussed in the previous chapter.
Adding the left-hand figure, 2, to the carried 6 we write down 8 and carry 4 to the right.

Step 3: For the hundreds place $2 \times 2 \times 1 + 3^2 + 2 \times 4 \times 2 + 3^2 + 2 \times 6 \times \bar{2} + \bar{1}^2 = 15$
And $4 + 1 = 5$ is written down and 5 is carried to the right.

Step 4: For the tens place $(3 \times 1 + 3 \times 2 + 1 \times 2) \times 2 = 22$
Adding the left-hand 2 to the carried 5 we write down 7 and carry 2 to the right.
So far we have

	2	3	1
	4	·3	2
	6	$\bar{1}$	$\bar{2}$

$$5\,_6\,8\,_4\,5\,_5\,7\,_2$$

Step 5: Finally, for the units place $1^2 + 2^2 + \bar{2}^2 = 9$
Hence the last two digits are 2 (carried) and 9.

> **(c) COMBINED ADDITION AND DIVISION**

Example 14   Divide $(2382 + 5617 + 7836 + 6195)$ by 7

We can record the sum, the working and the solution as follows.

	$\bar{1}$	1	3	
2	3	$_2$8	$_4$2	
$_1$5	6	1	$_1$7	
$_1$7	$_1$8	$_1$3	$_1$6	
$_1$6	1	$_1$9	$_1$5	
Answer 3	1	4	7 R1 $= 3147\frac{1}{7}$	

Working from left to right the steps are:

Step 1: Starting from the top of the first column $2 + 5 = 7$
This gives us one 7, which we note, putting a 1 just below and to the left of the 5.
Step 2: Continuing to work down the first column we next encounter a 7, which we register by writing a 1 at the left-hand foot of the 7.
Step 3: Continuing to work down the same column we next encounter a 6, which is 1 short of 7. We register a 1 at the left-hand foot of the 6 and the residue is –1. This –1 becomes –10 in the next column, and if we wish to write it down we can place it on top of a line drawn for this purpose above the sum; putting it in the carried position we need only write down $\bar{1}$.
Step 4: $\overline{10} + 3 + 6 = \bar{1}$,
and $\bar{1} + 8 = 7$ hence put 1 at the foot of the 8 and on its left.
Step 5: Continuing down the column, 1 is left, which can be placed above the line. So far we have:

$$
\begin{array}{cccc}
\overline{1} & 1 & & \\
\hline
2 & 3 & 8 & 2 \\
{}_1 5 & 6 & 1 & 7 \\
{}_1 7 & {}_1 8 & 3 & 6 \\
{}_1 6 & 1 & 9 & 5 \\
\hline
\end{array}
$$

Casting two 7's out of the 18 (= 10 + 8), we record 2 and hold 4 mentally

Step 6 : $4 + 1 = 5$ and $5 + 3 = 8$

Casting out 7 we record 1 and hold 1 mentally.

Step 7 : Casting 7 out of 9 and recording the 1 we have 2; add the 1 which was held over to make 3, and place this above the line, where it ranks as 30 (standing in the tens' position).

Step 8 : From 32 take four 7's, record the 4 and hold the remaining 4 (=32-28).

Step 9 : We record 1 for the next 7 and continue to hold 4.

Step 10: Since 6 is 1 short of 7 we record 1 and hold 4-1 = 3 mentally.

Step 11: $3 + 5 = 8$

We record 1, and the 1 remaining appears in the answer as $\frac{1}{7}$ (i.e. one not yet divided by 7).

Step 12: It now simply remains to add up the number of 7's cast out in each column, recording each total below its column, and there is the answer!

In this example each such total is a single digit and so no carrying is required- in fact we avoid such final carrying by taking a $\overline{1}$ to the right after passing down the first column.

The next example shows how the method can be adapted to deal with division by more than a single digit.

Example 15   Evaluate $(9245 + 3678 + 5937 + 381) \div 643$

Following our customary procedure, let us see what the completed calculation looks like after it has been written down:

$$
6^{43}\left|
\begin{array}{cccc}
\overline{1} & \overline{2} & & \overline{7} \\
\hline
{}^1 9 & {}^3 2 & 4 & 5 \\
{}^1 3 & {}^1 6 & 7 & 8 \\
{}^1 5 & {}^1 9 & 3 & 7 \\
0 & {}^1 3 & 8 & 1 \\
\hline
3 & 0 & & \overline{49} \\
\end{array}
\right.
$$

Answer 29 $\frac{594}{643}$

Note that the last two columns deal with the remainder portion- as many columns as there are flag digits; the dotted partition serves to remind us that this is so.

The first column on the left is worked through as if we were dividing by 6.
The details of these and other steps are as follows.

Step 1: Cast 6 out of the top left-hand 9. Record 1 and hold 3 mentally.

Step 2: $3 + 3 = 6$, record 1.

Step 3: The '5' being the last digit in this column, we record 1 and carry $\bar{1}$ to the right—possibly exercising our option of recording it, which is done above the line.

Step 4: For each 6 cast out in the first column we need to cast out a 4 in the second column. Hence, 3 being the first column total, we multiply it by 4 (the first flag-digit), giving us $3.4 = 12$ to be cast out—i.e. we have a contribution of -12 to add to the other figures in the second column (prior to recording any casting out).

Step 5: To this $\overline{12}$ we add $\bar{1}2$, giving $\overline{20}$.

Casting out three 6's we record $\bar{3}$ and hold $\bar{2}$ mentally.

Step 6: We record 1 for the next 6 and continue to hold $\bar{2}$ mentally.

Step 7: $9 + \bar{2} = 7$, record 1 and hold 1

Step 8: $3 + 1 = 4$.

Casting out 6 we record 1 and carry $\bar{2}$ to the right, recording it above the line.

So far we have:

		$\bar{1}$	$\bar{2}$		
		$^1 9$	$^3 2$	4	5
		$^1 3$	$^1 6$	7	8
6^{43}		$^1 5$	$^1 9$	3	7
		0	$^1 3$	8	1
		3	0		

Note that the total number of figures cast out in the first column is 3 and in the second column it is zero.

Step 9 : We perform a crosswise multiplication of the two flag-digits (4 and 3) by the two answer digits so far obtained (3 and 0): $4.0 + 3.3 = 9$

Step 10: This 9 is subtracted from the $\bar{2}4$ $(= \overline{20} + 4)$ at the head of the remainder section, leaving $\overline{25}$.

Step 11: There being no casting out of 6 in the remainder section we now proceed to total the figures left in the third column:
$\overline{25} + 7 + 3 + 8 = \bar{7}$.
Record the $\bar{7}$ above the line.

Step 12: It may have been observed that we have so far made use of two of the three parts of a "vertically and crosswise" multiplication of 43 (the flag-digits) and 30 (the solution digits). We subtracted 3.4 from column 2 and $0.4 + 3.3 = 9$ from column 3.
We now complete the pattern by subtracting $3.0 = 0$ from the last column.
We now need to total the last column: $1 + 7 + 8 + 5 - 70 = -49$
This last figure is the remainder and the answer is 30 remainder 49,

Or $30\frac{-49}{643}$ or $29\frac{594}{643}$.

A further example may help to clarify matters.
The few examples of this chapter serve as illustrations of what is evidently a much larger and fascinating topic- especially when bringing to bear the full range of sutras.

Example 16   Obtain the first five figures of $(96413.27 + 8347 + 96419.8) \div 8324$

We can write the question, working and solution down thus:

$$
\begin{array}{c}
\qquad\qquad 2\quad 2\quad 7\quad 9 \\
\hline
{}_1 9\ \ {}_2 6\ \ {}_0 4\ \ {}_6 1\ \ {}_7 3\ \ .2\ \ 7 \\
{}_1 8\ \ {}_1 3\ \ {}_0 4\ \ {}_1 7 \\
{}_1 9\ \ {}_1 6\ \ {}_0 4\ \ {}_0 1\ \ {}_0 9\ \ .8 \\
\hline
2\ \ \ 4\ \ .1\ \ 6\ \ \ 8 \\
\end{array}
$$

8^{324} (divisor, left of the bar)

Solution

The next digit leads us to round up the 8 to 9.
Alternatively, by working with negative digits we can keep to lower digits.
The above example could then be written thus:

$$
\begin{array}{c}
\qquad\qquad 2\quad 2\quad \bar{1}\quad \bar{2} \\
\hline
{}_1 9\ \ {}_2 6\ \ {}_1 4\ \ 1\ \ 3\ \ .2\ \ 7 \\
{}_1 8\ \ 3\ \ 4\ \ 7 \\
{}_1 9\ \ {}_1 6\ \ {}_1 4\ \ {}_{\bar 3} 1\ \ 9\ \ .8 \\
\hline
2\ \ 4\ \ .2\ \ \bar{3}\ \ \bar{1} \\
\end{array}
$$

8^{324}

Solution $2\ \ 4\ \ .1\ \ 6\ \ 9$

Exercises

(a) Sums of Products

First method

1) $7\times4 + 3\times2 + 9\times4 = 70$
2) $8\times3 + 7\times9 + 11\times12 = 219$
3) $2\times4 + 5\times7 + 3\times7 = 64$
4) $8\times2 + 6\times2 + 8\times1 = 36$
5) $2\times9 + 3\times4 + 2\times8 = 46$
6) $1\times4 + 3\times7 + 2\times5 = 35$
7) $7\times9 + 4\times8 + 7\times6 = 137$
8) $7\times6 + 8\times3 + 9\times5 = 111$
9) $2\times9 + 4\times3 + 7\times3 = 51$
10) $2\times11 + 12\times13 + 2\times5 = 188$

Second method   1) $22\times42 + 81\times92 = 8376$

2) $41\times54 + 90\times21 = 4104$

3) $24\times15 + 72\times39 = 3168$

4) $23\times17 + 37\times41 = 1908$

5) $15\times12 + 17\times24 = 588$

6) $423\times149 + 424\times627 = 328875$

7) $203\times249 + 372\times679 + 124\times763 = 397747$

8) $142\times279 + 652\times842 + 383\times436 = 755590$

9) $292\times420 + 708\times107 + 271\times586 = 357202$

10) $2473\times4218 + 4712\times1092 + 2401\times4297 = 52893715$

Third method   1) $8\times9 + 6\times7 = 114$

2) $4\times9 + 7\times7 = 81$

3) $8\times6 + 7\times6 = 90$

4) $16\times18 + 15\times19 + 20\times17 = 913$

5) $12\times14 + 16\times11 + 15\times19 = 629$

6) $12\times17 + 14\times18 + 14\times11 = 610$

7) $82\times42 + 16\times98 + 18\times52 = 5948$

8) $76\times47 + 82\times17 + 14\times83 = 6128$

9) $5883\times2681 + 2127\times8792 = 34472907$

Note that 88 and 12 are complementary (both middle digits)

10) $4682\times4124 + 2321\times2127 = 24245335$

Fourth method   1) $7\times4892 + 3\times5291 = 50117$

2) $5\times7829 + 8\times7112 = 96041$

3) $2\times2809 + 7\times3421 = 5618$

4) $5\times6812 + 3\times3431 = 44353$

5) $9\times7854 + 3\times9218 = 98340$

(b) Sums of Squares   1) $24^2 + 32^2 + 57^2 = 4849$

2) $59^2 + 42^2 + 71^2 = 10286$

3) $24^2 + 73^2 + 27^2 = 6634$

4) $242^2 + 821^2 + 584^2 = 1073661$

5) $281^2 + 524^2 + 129^2 = 370178$

6) $842^2 + 729^2 + 243^2 = 1299454$

7) $2431^2 + 2841^2 + 8243^2 = 81928091$

8) $2410^2 + 5314^2 + 3412^2 = 45688440$

9) $4123^2 + 7131^2 + 6111^2 = 105194511$

10) $3141^2 + 3117^2 + 7224^2 = 71767746$

(c) Combined Additions and Divisions

1) $(7429 + 3123 + 5149 + 3112) \div 6 = 3135.5$

2) $(2182 + 4621 + 8112 + 6127) \div 9 = 2338$

3) $(4118 + 2571 + 2650 + 2312) \div 8 = 1456.375$

4) $(2811 + 4628 + 3241 + 1127) \div 46 = 256.67391$

5) $(1421 + 2271 + 7112 + 6341) \div 32 = 535.78125$

6) $(4201 + 3621 + 4200 + 2234) \div 53 = 268.98113$
7)* $(27891 + 31172 + 11273 + 81127) \div 112 = 1352.3482$
8) $(14789 + 32461 + 82921 + 74433) \div 349 = 586.25788$
9) $(24628.432 + 82963.012 + 94236.7) \div 478 = 422.23461$
10)* $(58929 + 20631 + 56211 + 92134) \div 243 = 937.88066$

* Hints: in 7) cast out elevens initially;
        in 10) since $243 = 3 \times 81$, divide by 81 and then by 3.

Chapter 3

EVALUATION OF DETERMINANTS

How determinants arise is indicated in Chapter 4, on simultaneous equations.

SECOND ORDER DETERMINANTS

Methods for mental evaluation of second order determinants, and more generally of sums of products, were given in Chapter 2. These methods are assumed in what follows.

THIRD ORDER DETERMINANTS

A third order determinant is a layout, or matrix of elements specifying nine multiplications and various additions and subtractions, leading finally to a single figure. Thus a third order determinant is a formula reducing $3^2 = 9$ figures to one.

Example 1 Evaluate $D = \begin{vmatrix} 2 & 6 & 7 \\ 5 & 4 & 8 \\ 3 & 1 & 9 \end{vmatrix}$

Let us separate the first row from the second and third by a partition:

$$D = \begin{vmatrix} 2 & 6 & 7 \\ 5 & 4 & 8 \\ 3 & 1 & 9 \end{vmatrix}$$

Three second-order determinants arise from the second and third rows:

$$\begin{vmatrix} 5 & 4 \\ 3 & 1 \end{vmatrix} = -7; \qquad \begin{vmatrix} 5 & 8 \\ 3 & 9 \end{vmatrix} = 21; \qquad \begin{vmatrix} 4 & 8 \\ 1 & 9 \end{vmatrix} = 28$$

The sign of the second of these is reversed (by calculating $8 \times 3 - 5 \times 9 = -21$, and the three results, −7, −21 and 28 are noted in a fourth row:

$$\begin{vmatrix} 2 & 6 & 7 \\ 5 & 4 & 8 \\ 3 & 1 & 9 \end{vmatrix}$$

$$-7 \quad -21 \quad 28$$

The first and fourth rows are now multiplied together VERTICALLY AND CROSSWISE, i.e. we take $\quad 2\times28 + 6\times(-21) + 7\times(-7) = -119$.

$\therefore \underline{D = -119}$

Example 2   Evaluate $E = \begin{vmatrix} -3 & 7 & 1 \\ 2 & 4 & 3 \\ 5 & 8 & -4 \end{vmatrix}$

$$-26 \quad +11 \quad +17$$

$\therefore E = \begin{vmatrix} -3 & 7 & 1 \\ 2 & 4 & 3 \\ 5 & 8 & -4 \end{vmatrix}$ $\underline{E = +277}$

Example 3   Evaluate $F = \begin{array}{c} 37 \\ -32 \\ 19 \end{array} \begin{vmatrix} 7 & -3 & 9 \\ 3 & 4 & 5 \\ -1 & 5 & 7 \end{vmatrix}$ $= \underline{270}$

Note that the partitioning has been applied in various places; in Example 3 it is by columns. The multiplication pattern ✕ (see Chapter 1) now becomes ⅄

Since interchanging rows and columns does not change a determinant, evidently the sutra VERTICALLY AND CROSSWISE needs to be understood as applying to both these patterns.

Note also that, of the three second-order determinants calculated after partitioning, that with a negative sign contains elements spaced one row (or column) apart; adjacent elements lead to a positive sign. This observation is useful in dealing with fourth and higher order determinants. The following example should make the point clear.

FOURTH ORDER DETERMINANTS

Partitioning centrally (either horizontally or vertically) we now proceed to evaluate second order determinants either side of the partition.

Example 4   Evaluate $D = \begin{vmatrix} 2 & 5 & 7 & 8 \\ 3 & 6 & 2 & 4 \\ 3 & 5 & 7 & 6 \\ 1 & 6 & 9 & 7 \end{vmatrix}$

$$
\begin{array}{cccccc}
+ & - & + & - & + & - \\
-3 & -17 & -16 & -32 & -28 & 12
\end{array}
$$

$$
= \begin{array}{|cccc|}
\hline
2 & 5 & 7 & 8 \\
3 & 6 & 2 & 4 \\
\hdashline
3 & 5 & 7 & 6 \\
1 & 6 & 9 & 7 \\
\hline
\end{array}
$$

$$
\begin{array}{cccccc}
13 & 20 & 15 & 3 & -1 & -5
\end{array}
$$

The second-order determinants above the partition are:

$\begin{vmatrix} 2 & 5 \\ 3 & 6 \end{vmatrix} = -3;$ $\begin{vmatrix} 2 & 7 \\ 3 & 2 \end{vmatrix} = -17;$ $\begin{vmatrix} 2 & 8 \\ 3 & 4 \end{vmatrix} = -16;$

$\begin{vmatrix} 5 & 7 \\ 6 & 2 \end{vmatrix} = -32;$ $\begin{vmatrix} 5 & 8 \\ 6 & 4 \end{vmatrix} = -28;$ $\begin{vmatrix} 7 & 8 \\ 2 & 4 \end{vmatrix} = 12;$

These are placed above the determinant in a row, the associated signs being placed above, "+" for adjacent columns, "–" for those spaced one column apart, "+" for those spaced two apart.

The second-order determinants below the partition are obtained in the same sequence:

$\begin{vmatrix} 3 & 5 \\ 1 & 6 \end{vmatrix} = 13;$ $\begin{vmatrix} 3 & 7 \\ 1 & 9 \end{vmatrix} = 20;$ $\begin{vmatrix} 3 & 6 \\ 1 & 7 \end{vmatrix} = 15;$

$\begin{vmatrix} 5 & 7 \\ 6 & 9 \end{vmatrix} = 3;$ $\begin{vmatrix} 5 & 6 \\ 6 & 7 \end{vmatrix} = -1;$ $\begin{vmatrix} 7 & 6 \\ 9 & 7 \end{vmatrix} = -5.$

The two outermost rows of figures are now multiplied together according to

the pattern: $\bowtie$

associating the relevant signs with them.
I.e. we have D = +(–3)(–5) – (–17)(–1) + (–16)(3) + . . . + (12)(13) = <u>186</u>

A pyramid layout of the second-order sub-determinants is sometimes useful:

$$
\begin{array}{l}
+ \\
- \\
+ \\
\end{array}
\quad
\begin{array}{ccc}
 & -16 & \\
-17 & & -28 \\
-3 & -32 & +12 \\
\end{array}
$$

$$
\begin{array}{|cccc|}
\hline
2 & 5 & 7 & 8 \\
3 & 6 & 2 & 4 \\
\hdashline
3 & 5 & 7 & 6 \\
1 & 6 & 9 & 7 \\
\hline
\end{array}
$$

$$
\begin{array}{cccc}
13 & & 3 & -5 \\
 & 20 & -1 & \\
 & 15 & &
\end{array}
$$

The first row above the determinant stems from adjacent elements,

viz: $\begin{vmatrix} 2 & 5 \\ 3 & 6 \end{vmatrix} = -3$; $\begin{vmatrix} 5 & 7 \\ 6 & 2 \end{vmatrix} = -32$; $\begin{vmatrix} 7 & 8 \\ 2 & 4 \end{vmatrix} = 12$;

These all have a positive sign associated with them, as indicated in the column of signs on the left.

The next row stems from elements spaced one column apart, namely:

$\begin{vmatrix} 2 & 7 \\ 3 & 2 \end{vmatrix} = -17$; $\begin{vmatrix} 5 & 8 \\ 6 & 4 \end{vmatrix} = -28$;

As the sign-column indicates, these have an associated minus sign.

At the top of the pyramid comes: $\begin{vmatrix} 2 & 8 \\ 3 & 4 \end{vmatrix} = -16$;

Coming from elements spaced two columns apart, this has a "+" sign associated with it.

Similarly with the lower inverted pyramid.

The pattern by which the elements of the pyramid are multiplied together can be summed thus:

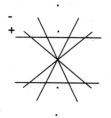 and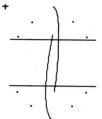

i.e. D = +(−3)(−5) + (12)(13) − (−17)(−1) − (−28)(20) + (−16)(3) + (−32)(15)
= 186, as before.

REDUCING THE ORDER OF A DETERMINANT BY ONE

The method of pivotal reduction is known to mathematicians, but not so widely used as it deserves.

Example 5 $D = \begin{vmatrix} 2 & 3 & 7 \\ 8 & 9 & 5 \\ 6 & 10 & 3 \end{vmatrix}$

Using the "2" as pivot we can write: $D = \frac{1}{2} \begin{vmatrix} \begin{vmatrix} 2 & 3 \\ 8 & 9 \end{vmatrix} & \begin{vmatrix} 2 & 7 \\ 8 & 5 \end{vmatrix} \\ \begin{vmatrix} 2 & 3 \\ 6 & 10 \end{vmatrix} & \begin{vmatrix} 2 & 7 \\ 6 & 3 \end{vmatrix} \end{vmatrix}$

$\therefore D = \frac{1}{2} \begin{vmatrix} -6 & -46 \\ 2 & -36 \end{vmatrix} = \underline{154}$

The procedure is that each element not in the same row or column as the pivot is replaced by a second order determinant. The latter is formed from the elements at the corners of a rectangle, of which the elements in question and the pivot constitute two opposite corners. There is also a division by a^{n-2}, where "a" denotes the pivot and "n" the order of the determinant before reduction.

In the following examples the element chosen as pivot is asterisked.

Example 6 $D = \begin{vmatrix} 5 & 3 & 4 \\ 6 & 3* & 2 \\ 3 & 5 & 7 \end{vmatrix} = \frac{1}{3} \begin{vmatrix} -3 & -6 \\ 21 & 11 \end{vmatrix} = \underline{31}$

Example 7 $D = \begin{vmatrix} 2* & 1 & 3 \\ 1 & 2 & 4 \\ 3 & 1 & 5 \end{vmatrix} = \frac{1}{2} \begin{vmatrix} 3 & 5 \\ -1 & 1 \end{vmatrix} = \underline{4}$

Example 8 $D = \begin{vmatrix} 6 & 1 & 3 & 2 \\ 5 & 0 & 3 & 4 \\ 1* & 3 & 4 & 2 \\ 5 & 1 & 2 & 3 \end{vmatrix} = \begin{vmatrix} 17 & 21 & 10 \\ 15 & 17 & 6 \\ -14 & -18 & -7 \end{vmatrix}$

The well-known rule, that a determinant is unchanged on adding one row or column to another, can usefully be applied here. Adding row 2 to row 3, and subtracting row 2 from row 1 we have:

$D = \begin{vmatrix} 2 & 4 & 4 \\ 15 & 17 & 6 \\ 1* & -1 & -1 \end{vmatrix} = \begin{vmatrix} -6 & -6 \\ -32 & -21 \end{vmatrix} = 6(21 - 32) = \underline{-66}$

REDUCTION OF A DETERMINANT BY TWO DEGREES

The foregoing method can be applied in succession, but there is a more direct way of achieving the same result. Instead of using a single element as pivot we can use a second order determinant as pivot, replacing elements in all other rows and columns by third order determinants.

Example 9 Given $D = \begin{vmatrix} 2 & 0 & 3 & 5 \\ 0 & 1 & 7 & 3 \\ 1 & 5 & 2 & 4 \\ 4 & 3 & 2 & 6 \end{vmatrix}$ by reducing the determinant to one of second order,

evaluate D.

Here we can take advantage of the two zeros and use $\begin{vmatrix} 2 & \overset{\bullet}{0} \\ 0 & 1 \end{vmatrix}$ as pivot.

The steps are:

$$D = \tfrac{1}{2}\begin{vmatrix} \begin{vmatrix} 2 & 0 & 3 \\ 0 & 1 & 7 \\ 1 & 5 & 2 \end{vmatrix} & \begin{vmatrix} 2 & 0 & 5 \\ 0 & 1 & 3 \\ 1 & 5 & 4 \end{vmatrix} \\ \begin{vmatrix} 2 & 0 & 3 \\ 0 & 1 & 7 \\ 4 & 3 & 2 \end{vmatrix} & \begin{vmatrix} 2 & 0 & 5 \\ 0 & 1 & 3 \\ 4 & 3 & 6 \end{vmatrix} \end{vmatrix} \qquad \ldots(1)$$

$$= \tfrac{1}{2}\begin{vmatrix} -69 & -27 \\ -50 & -26 \end{vmatrix} = \underline{222}$$

Note that the divisor = $(\text{pivot})^{n-3}$, where "n" is the order of the original determinant.

We can check the method by two successive pivotal reductions.
Using the 1 in row 2, column 2 as pivot, we have:

$$D = \begin{vmatrix} \begin{vmatrix} 2 & 0 \\ 2 & 1 \end{vmatrix}^{*} & \begin{vmatrix} 0 & 3 \\ 1 & 7 \end{vmatrix} & \begin{vmatrix} 0 & 5 \\ 1 & 3 \end{vmatrix} \\ \begin{vmatrix} 0 & 1 \\ 1 & 5 \end{vmatrix} & \begin{vmatrix} 1 & 7 \\ 5 & 2 \end{vmatrix} & \begin{vmatrix} 1 & 3 \\ 5 & 4 \end{vmatrix} \\ \begin{vmatrix} 0 & 1 \\ 4 & 3 \end{vmatrix} & \begin{vmatrix} 1 & 7 \\ 3 & 2 \end{vmatrix} & \begin{vmatrix} 1 & 3 \\ 3 & 6 \end{vmatrix} \end{vmatrix}$$

And now applying a second pivotal reduction using the asterisked term:

$$D = \tfrac{1}{2}\begin{vmatrix} \begin{vmatrix} \begin{vmatrix} 2 & 0 \\ 0 & 1 \end{vmatrix} & \begin{vmatrix} 0 & 3 \\ 1 & 7 \end{vmatrix} \\ \begin{vmatrix} 0 & 1 \\ 1 & 5 \end{vmatrix} & \begin{vmatrix} 1 & 7 \\ 5 & 2 \end{vmatrix} \end{vmatrix} & \begin{vmatrix} \begin{vmatrix} 2 & 0 \\ 0 & 1 \end{vmatrix} & \begin{vmatrix} 0 & 5 \\ 1 & 3 \end{vmatrix} \\ \begin{vmatrix} 0 & 1 \\ 1 & 5 \end{vmatrix} & \begin{vmatrix} 1 & 3 \\ 5 & 4 \end{vmatrix} \end{vmatrix} \\ \begin{vmatrix} \begin{vmatrix} 2 & 0 \\ 0 & 1 \end{vmatrix} & \begin{vmatrix} 0 & 3 \\ 1 & 7 \end{vmatrix} \\ \begin{vmatrix} 0 & 1 \\ 4 & 3 \end{vmatrix} & \begin{vmatrix} 1 & 7 \\ 3 & 2 \end{vmatrix} \end{vmatrix} & \begin{vmatrix} \begin{vmatrix} 2 & 0 \\ 0 & 1 \end{vmatrix} & \begin{vmatrix} 0 & 5 \\ 1 & 3 \end{vmatrix} \\ \begin{vmatrix} 0 & 1 \\ 4 & 3 \end{vmatrix} & \begin{vmatrix} 1 & 3 \\ 3 & 6 \end{vmatrix} \end{vmatrix} \end{vmatrix} \qquad \ldots(2)$$

If we reduce the third order determinants in Expression (1) to second-order determinants, using the central "1" as pivot, we arrive at Expression (2), which shows the equivalence of the two procedures.

Note that the third-order determinants consist of elements in the same row (or column) as the element being replaced, and in the same columns (or rows) as the second-order pivot.

In general there is no advantage in direct reduction of a determinant by two orders; the method involves far more multiplications than two successive pivotal reductions (using a single element as pivot). Under special conditions there may be little difference between the number of multiplications required by the two methods, and direct reduction by two orders can be worthwhile. Those special conditions are met in the above example: two zeros at opposite corners of the 2 × 2 pivot, and (for preference) the other two elements either equal to unity or simple multiples of unity. Under these conditions mental evaluation of the third order determinants involved is facilitated.

Note that if there are three zeros in the pivot the original determinant vanishes. Again, a zero pivot is not acceptable, divisions by zero being disallowed; this disposes of the case of a pivot with zeros in one row or column.

EVALUATING A DETERMINANT BY ROW AND COLUMN OPERATIONS

This method is well-known. In the following examples, rows are denoted by R_1, R_2, etc., and columns by C_1, C_2 etc.
The operation of subtracting Row 2 from Row 3, e.g., is denoted by $R_3 - R_2$.

A spacious layout and an orderly (here clockwise) sequence for replacing eliminated or modified elements, aids clarity during the working.

Example 10 Evaluate $D = \begin{vmatrix} 52 & 31 & 17 \\ 5 & 3 & 4 \\ 16 & 8 & 3 \end{vmatrix}$

Working

$$\begin{vmatrix} \underset{1}{+\ \cancel{52}} & \overset{-3}{\cancel{31}}_{\ 0} & \underset{10}{17} \\ {}_{-7}\ \cancel{5}\ _0 & \overset{-5}{\cancel{3}}_{\ -3} & 4\ _7 \\ {}_7\ \cancel{16} & \overset{2}{\cancel{8}} & 3 \end{vmatrix}$$

STEPS
1) $C_1 - 3C_3$
2) $C_2 - 2C_3$
3) $R_2 + R_3$
4) $R_1 - R_2$

Note that four of the eight positions around the figure are here used in sequence to record the results of successive row or column operations.

This gives: $D = \begin{vmatrix} 1 & 0 & 10 \\ 0 & -3 & 7 \\ 7 & 2 & 3 \end{vmatrix}$

$\therefore$ D = <u>187</u>, on expansion by first row (say).

The Vedic ideal is mental calculation. Just as it is possible to play chess blindfolded, so it should be possible to perform all these calculations mentally- with fifth-order determinants, let alone the third and fourth order.

| EXTRACTION OF ELEMENTS | (Expansion by individual elements)

Expansion of a determinant is normally done by a row or column. But these are not the only options, as will now be shown.

| **Extraction of one Element** |

Example 11 $D = \begin{vmatrix} 3 & 7 & 10 \\ 4 & 129 & 3 \\ 4 & 2 & 5 \end{vmatrix}$

The 129 is an exceptionally large element. We may wish to extract it, giving:

$$D = \begin{vmatrix} 3 & 7 & 10 \\ 4 & 0 & 3 \\ 4 & 2 & 5 \end{vmatrix} + 129 \begin{vmatrix} 3 & 10 \\ 4 & 5 \end{vmatrix}$$

This can be seen to hold good on considering expansion by second row (or second column).

The procedure is: first eliminate the element, then everything else but that element and its expansion (the terms that go with it). The sutra BY ALTERNATE ELIMINATION AND RETENTION is in evidence here.

Alternatively, we may choose to extract 127 from the 129, leaving 2 in that position.
We then have:

$$D = \begin{vmatrix} 3 & 7 & 10 \\ 4 & 2 & 3 \\ 4 & 2 & 5 \end{vmatrix} + 127 \begin{vmatrix} 3 & 10 \\ 4 & 5 \end{vmatrix}$$

Subtracting the second row of the left-hand determinant from the third and expanding by the third row gives:

$$D = 2 \begin{vmatrix} 3 & 7 \\ 4 & 2 \end{vmatrix} + 127 \begin{vmatrix} 3 & 10 \\ 4 & 5 \end{vmatrix} = -44 - 3175 = \underline{-3219}$$

Extraction of Two Elements

Example 12 $E = \begin{vmatrix} 2 & 5 & 87 \\ 2 & 93 & 3 \\ 4 & 6 & 5 \end{vmatrix} = \begin{vmatrix} 2 & 5 & 0 \\ 2 & 0 & 3 \\ 4 & 6 & 5 \end{vmatrix} + 93 \begin{vmatrix} 2 & 0 \\ 4 & 5 \end{vmatrix} + 87 \begin{vmatrix} 2 & 0 \\ 4 & 6 \end{vmatrix} + 87\%93\%4$

Here the elements 87 and 93 being unusually large, they have been extracted. The procedure is, first eliminate both elements, then expand one eliminating the other, and vice-versa; then expand both together.

As explained above, the "expansion" of an element is a brief reference to the terms that go with that element.

This can be proved using the rule for extracting one element twice in succession. Note that for each element extracted the number of determinants is doubled.

Example 13 $D = \begin{vmatrix} 153 & 3 & 4 & 5 \\ 8 & 149 & 6 & 7 \\ 11 & 7 & 203 & 8 \\ 5 & 10 & 9 & 315 \end{vmatrix}$

$= \begin{vmatrix} 0 & 3 & 4 & 5 \\ 8 & 0 & 6 & 7 \\ 11 & 7 & 203 & 8 \\ 5 & 10 & 9 & 315 \end{vmatrix} + 153 \begin{vmatrix} 0 & 6 & 7 \\ 7 & 203 & 8 \\ 0 & 9 & 315 \end{vmatrix} + 149 \begin{vmatrix} 0 & 4 & 5 \\ 11 & 203 & 8 \\ 5 & 9 & 315 \end{vmatrix}$

$+ 149\%153 \begin{vmatrix} 203 & 8 \\ 9 & 315 \end{vmatrix}$

Extraction of Three Elements

Example 14 $D = \begin{vmatrix} 153 & 3 & 4 & 5 \\ 8 & 149 & 6 & 7 \\ 11 & 7 & 203 & 8 \\ 5 & 10 & 9 & 315 \end{vmatrix}$

$= \begin{vmatrix} 0 & 3 & 4 & 5 \\ 8 & 0 & 6 & 7 \\ 11 & 7 & 0 & 8 \\ 5 & 10 & 9 & 315 \end{vmatrix} + 203 \begin{vmatrix} 0 & 3 & 5 \\ 8 & 0 & 7 \\ 5 & 10 & 315 \end{vmatrix} + 153 \begin{vmatrix} 0 & 6 & 7 \\ 7 & 0 & 8 \\ 10 & 9 & 315 \end{vmatrix}$

$+ 149 \begin{vmatrix} 0 & 4 & 5 \\ 11 & 0 & 8 \\ 5 & 9 & 315 \end{vmatrix} + 153\%203 \begin{vmatrix} 0 & 7 \\ 10 & 315 \end{vmatrix} + 149\%203 \begin{vmatrix} 0 & 5 \\ 5 & 315 \end{vmatrix}$

$$+ \; 149\,\%153 \begin{vmatrix} 0 & 8 \\ 9 & 315 \end{vmatrix} + 149\%153 \; \%203 \; \%315$$

The rule employed is as follows:
First we eliminate the three elements; then we expand one eliminating the other two, going through the three cases; then we expand two at a time, eliminating the third- and this can be done three ways; then we expand three at a time.

Example 15 $D = \begin{vmatrix} L-2 & 5 & 7 \\ 4 & L-3 & 9 \\ 6 & 8 & L-4 \end{vmatrix}$

Suppose we choose to extract the three L's.

$$\text{Then } D = \begin{vmatrix} -2 & 5 & 7 \\ 4 & -3 & 9 \\ 6 & 8 & -4 \end{vmatrix} + L \left\{ \begin{vmatrix} -3 & 9 \\ 8 & -4 \end{vmatrix} + \begin{vmatrix} -2 & 7 \\ 6 & -4 \end{vmatrix} + \begin{vmatrix} -2 & 5 \\ 4 & -3 \end{vmatrix} \right\}$$

$$+ \; L^2(-2-3-4) + L^3$$

$\therefore \; \underline{D = 800 - 108L - 9L^2 + L^3}$

The eigen-value problem concerns finding the roots of D=0, i.e. of the above cubic in this case, which can be done using the methods of Chapter 12.

A SPECIAL CASE

The **VERTICALLY AND CROSSWISE** sutra is well-adapted to handling general cases rather than special cases. By way of contrast a sutra will now be used where the reverse appears to apply.

This special case comes under the sutra, **WHEN THE SAMUCCAYA IS THE SAME, THAT SAMUCCAYA IS ZERO.** In this context the Sanskrit word "samuccaya" can be understood to mean "total"—although Tirthaji does not always use it in this way (see "Vedic Mathematics").

Before dealing with examples of this special case let us consider the reasoning behind it with the aid of two simple examples.

Example 16 $D = \begin{vmatrix} 3 & 2 & 20 \\ 5 & 3 & 17 \\ 6 & 8 & 11 \end{vmatrix} = \begin{vmatrix} 3 & 2 & 25 \\ 5 & 3 & 25 \\ 6 & 8 & 25 \end{vmatrix}$

Here the first and second columns have been added to the third.

$$\therefore D = 25 \begin{vmatrix} 3 & 2 & 1 \\ 5 & 3 & 1 \\ 6 & 8 & 1 \end{vmatrix}$$

i.e. the row-totals being the same in each case, that total is a factor of the determinant.

Example 17 $D = \begin{vmatrix} 9+L & 7 & 5 \\ 7 & 5+L & 9 \\ 5 & 9 & 7+L \end{vmatrix} = (21+L) \begin{vmatrix} 1 & 1 & 1 \\ 7 & 5+L & 9 \\ 5 & 9 & 7+L \end{vmatrix}$

In the second example each column (or row) contains the same elements, differing only in sequence, and hence the column-total (which equals the row-total) is a factor,
i.e. $9 + 7 + 5 + L = 21 + L$ is a factor.

These preliminaries lead to the following applications of the sutra WHEN THE SAMUCCAYA IS THE SAME, THAT SAMUCCAYA IS ZERO.

Example 18   Find a root of $D = \begin{vmatrix} 3+L & 9 & 7 \\ 7+L & 3 & 9 \\ 9 & 7 & 3+L \end{vmatrix} = 0$

The same elements, 3, 9 and 7 occur in each row, in differing sequences
$\therefore 3 + 9 + 7 + L = 0$
I.e. $\underline{L = -19}$ is one solution.

Example 19   Find one root of $D = \begin{vmatrix} L+1 & 2 & 3 & 4 \\ 2 & L+3 & 1 & 4 \\ 3 & 1 & L+4 & 2 \\ 1 & 4 & 2 & L+3 \end{vmatrix} = 0$

The total in each row is $L + 1 + 2 + 3 + 4$, therefore the sutra applies, and $L + 10 = 0$ gives one solution.

Example 20   Find a root of $E = \begin{vmatrix} L+3 & 2 & 2L-5 & 4 \\ 2 & L-5 & 4 & 2L+3 \\ 4 & 2L+3 & L+2 & -5 \\ 2L-5 & 4 & 3 & L+2 \end{vmatrix} = 0$

Here the total in each row (and each column) is the same, viz. $3L + 4$.
$\therefore L = -\frac{4}{3}$ is one solution.

Example 21 Find one or more roots of $F = \begin{vmatrix} L-6 & L^2+3 & 2 \\ 3 & L+2 & L^2-6 \\ 2 & L^2-6 & L+3 \end{vmatrix} = 0$

The total of each row being the same, i.e. $L^2 + L - 1$, it follows that
$L^2 + L - 1 = 0$

I.e. $L = \dfrac{-(1+\sqrt{5})}{2}$ and $L = \dfrac{-1+\sqrt{5}}{2}$ are two of the six roots.

Exercises

Evaluate each of the following determinants by two methods:

1) $D = \begin{vmatrix} 3 & 2 & 1 \\ 2 & 7 & 1 \\ 5 & 6 & 5 \end{vmatrix}$ (D = 54)

2) $D = \begin{vmatrix} 4 & 1 & 8 \\ 6 & 2 & 3 \\ 2 & 1 & 0 \end{vmatrix}$ (D = 10)

3) $D = \begin{vmatrix} 2 & 3 & 2 \\ 4 & 7 & 3 \\ 2 & 1 & 3 \end{vmatrix}$ (D = –2)

4) $D = \begin{vmatrix} 4 & 5 & 5 & 7 \\ 7 & 6 & 8 & 8 \\ 2 & 2 & 2 & 3 \\ 3 & 4 & 9 & 4 \end{vmatrix}$ (D = 9)

5) $D = \begin{vmatrix} 2 & 9 & 4 & 3 \\ 5 & 6 & 7 & 1 \\ 2 & 4 & 3 & 5 \\ 1 & 4 & 3 & 2 \end{vmatrix}$ (D = 145)

6) $D = \begin{vmatrix} 3 & 2 & 4 & 6 \\ 2 & 5 & 7 & 1 \\ 2 & 4 & 1 & 3 \\ 5 & 4 & 6 & 2 \end{vmatrix}$ (D = 414)

Extract the appropriate elements from:

7) $D = \begin{vmatrix} 3 & 4 & 2 \\ 5 & 98 & 7 \\ 2 & 1 & 4 \end{vmatrix}$

8) $D = \begin{vmatrix} 5 & 6 & 7 \\ 8 & 9 & 4 \\ 2 & 88 & 2 \end{vmatrix}$

9) $D = \begin{vmatrix} 4 & 67 & 4 \\ 2 & 3 & 42 \\ 1 & 7 & 8 \end{vmatrix}$

10) $D = \begin{vmatrix} 58 & 2 & 3 \\ 4 & 2 & 1 \\ 6 & 7 & 67 \end{vmatrix}$

11) $D = \begin{vmatrix} 48 & 2 & 3 & 2 \\ 6 & 99 & 2 & 5 \\ 3 & 2 & 111 & 4 \\ 4 & 3 & 3 & 120 \end{vmatrix}$

12) $D = \begin{vmatrix} 3 & 45 & 6 & 7 \\ 98 & 2 & 1 & 7 \\ 4 & 8 & 97 & 6 \\ 4 & 2 & 3 & 4 \end{vmatrix}$

Evaluate the following 'special case' determinant:

13) $D = \begin{vmatrix} 4 & 9 & 11 \\ 6 & 2 & 16 \\ 9 & 2 & 13 \end{vmatrix}$ (D = 264)

Find a solution to the following 'special case' equations:

14) $D = \begin{vmatrix} 7+L & 2 & 5 \\ 6 & 3+L & 5 \\ 3 & 4 & 7+L \end{vmatrix} = 0$ (L = -14)

15) $D = \begin{vmatrix} L^2+3 & -3L & -2 \\ 4 & L^2-3 & -3L \\ -3L+6 & 2 & L^2-7 \end{vmatrix} = 0$ $(L = \frac{3\pm\sqrt{5}}{2})$

Chapter 4

THE SOLUTION OF SIMULTANEOUS LINEAR EQUATIONS

A method for the solution of two equations in two unknowns is given in "Vedic Mathematics", and also some ways of solving three equations in three unknowns by procedures familiar to western mathematicians. Further methods are given here which draw on the Vedic system of sutras for solving equations in two, three or more unknowns. The VERTICALLY AND CROSSWISE sutra is in evidence here, although the layout employed sometimes lays the 'vertical' down sideways.

In "Vedic Mathematics" the following method is given for solving two equations in two unknowns.

Example 1 $2x + 3y = 8$ ...(1)
$x + 4y = 9$ ...(2)

$\square$ $x = \frac{3\%9 - 4\%8}{3\%1 - 2\%4} = \frac{-5}{-5} = 1$

$y = \frac{8\%1 - 2\%9}{3\%1 - 2\%4} = \frac{-10}{-5} = 2$

There is a definite similarity in this to Kramer's Rule, which yields:

$$x = \dfrac{-\begin{vmatrix} 3 & 8 \\ 4 & 9 \end{vmatrix}}{\begin{vmatrix} 2 & 3 \\ 1 & 4 \end{vmatrix}} = \frac{5}{5} = 1 \qquad y = \dfrac{\begin{vmatrix} 2 & 8 \\ 1 & 9 \end{vmatrix}}{\begin{vmatrix} 2 & 3 \\ 1 & 4 \end{vmatrix}} = 2$$

In these methods use is made of three 2 by 2 determinants, arising from the two rows of three coefficients occurring in Equations (1) and (2), namely:

$$\begin{pmatrix} 2 & 3 & 8 \\ 1 & 4 & 9 \end{pmatrix}$$
...(3)

This gives the sense of dealing with two rows of a third order determinant. But if we are dealing with a third order determinant, what is the missing row?

The three second order determinants give, on sorting out the signs, x.D, y.D and D,

where $D = \begin{vmatrix} -2 & 3 \\ 1 & 4 \end{vmatrix}$

Furthermore, x.D is given by the determinant arising on missing out the first column of the matrix (3), –y.D from missing out the second column and –D from missing out the third. This leads us to consider the third order determinant:

$$\det A = \begin{vmatrix} x.D & y.D & -D \\ 2 & 3 & 8 \\ 1 & 4 & 9 \end{vmatrix} \qquad\qquad\qquad ...(4)$$

On evaluating det A, the coefficient of x.D yields the value of x.D, the coefficient of y.D yields the value of y.D, and the coefficient of D yields the value of D.

i.e.   det A = –5x.D – 10y.D – 5D,
hence   D = –5, x.D = –5 and y.D = –10
Whence x = 1 and y = 2

The procedure is found to hold good for solving 'n' equations in 'n' unknowns. This leads to various methods of solving n equations in n unknowns, corresponding to various methods of evaluating determinants- for which see Chapter 3.

Returning to our present example, note that if the right-hand-side terms of Equations (1) and (2) were brought to the left, the need to associate a minus sign with D would vanish, and we would have:

$$\det A = -\begin{vmatrix} x.D & y.D & D \\ 2 & 3 & -8 \\ 1 & 4 & -9 \end{vmatrix}$$

Proceeding with this example for the fun of it, and also because it illustrates possibilities which are perhaps of more practical use in higher order examples, we have, on extracting the common factor D from the first row:

$$\det A = -D\begin{vmatrix} x & y & 1 \\ 2 & 3 & -8 \\ 1 & 4 & -9 \end{vmatrix}$$

Following the notation of Chapter 3, denote the three rows by R_1, R_2 and R_3 and the columns by C_1, C_2 and C_3 and let the increase of R_2 by the addition of R_1 eight times be represented by $R_2 + 8R_1$.
Then the two steps 1) $R_3 + R_1 - R_2$ and 2) $R_2 + 8R_1$, yield:

$$\det A = -D\begin{vmatrix} x & y & 1 \\ 2+8x & 3+8y & 0 \\ x-1 & 1+y & 0. \end{vmatrix}$$

Expanding by the third column we have:

$$\det A = -D \begin{vmatrix} 2+8x & 3+8y \\ -1+x & 1+y \end{vmatrix}$$

And now multiplying out, we have: $\det A = -\{(2+3)D + (8-3)x.D + (2+8)y.D\}$.

The minus sign outside the curly brackets can now be ignored- it does not affect the answer.

Bearing in mind that the coefficients of D, x.D and y.D give the respective values of D, x.D and y.D, we have:

$D = 5$, x.D $= 5$ and y.D $= 10$, giving x $= 1$ and y $= 2$

Non-linear terms, such as x.y, invariably cancel out, and can be ignored. This can be verified by expanding the original determinant by its first row.

In the next example, for three equations in three unknowns, the top row of the fourth order determinant need not be shown, but can be understood implicitly. A central partition is used to aid calculation of the four third order determinants arising on expanding by the (implicit) top row.

Note that, just as it is invalid to divide by zero, so it is invalid to make the coefficient of D zero (i.e. the top right-hand element of the determinant). This is a restriction on the procedure, but not one likely to cause any practical difficulty.

Example 2   To be solved: $4x - 2y + 3z = 8$
$$2x - 3y - z = 1$$
$$3x + 2y - z = 3$$

The solution can be set out as follows:

$$\begin{array}{c} -8 \\ -14 \\ 13 \end{array} \begin{vmatrix} xD & yD & zD & -D \\ 4 & -2 & 3 & 8 \\ 2 & -3 & 1 & 1 \\ 3 & 2 & -1 & 3 \end{vmatrix} \begin{array}{c} -5 \\ -17 \\ 4 \end{array}$$

Explanation
The four columns of coefficients are partitioned centrally, and the three 2 by 2 determinants to the left of the partition are evaluated:

$$\begin{vmatrix} 4 & -2 \\ 2 & -3 \end{vmatrix} = -8; \qquad -\begin{vmatrix} 4 & -2 \\ 3 & 2 \end{vmatrix} = -14; \qquad \begin{vmatrix} 2 & -3 \\ 3 & 2 \end{vmatrix} = 13$$

These are written in the column on the left, and the column on the extreme right is similarly evaluated from:

$$\begin{vmatrix} 3 & 8 \\ 1 & 1 \end{vmatrix} = -5; \qquad -\begin{vmatrix} 3 & 8 \\ 1 & 3 \end{vmatrix} = -17; \qquad etc.$$

The similarity between this method and Kramer's Rule is in evidence at the next step. Covering the fourth column, we evaluate the 3 by 3 determinant of the remaining three columns using a minor variant of the cross-product procedure encountered in Chapter 1.

We have
$$
\begin{array}{cc}
-8 & 3 \\
-14 & 1 \\
13 & -1
\end{array}
$$
i.e.
$$
\begin{array}{l}
(-8) \times (-1) \\
+(-14) \times\ 1 \\
\underline{+\ \ 13\ \times\ \ 3} \\
\text{Total} = D = 33
\end{array}
$$

Then to evaluate x.D cover the first column and evaluate the 3 by 3 determinant of the remaining three columns using:

$$
\begin{array}{cc}
-2 & -5 \\
-3 & -17 \\
2 & 4
\end{array}
$$
i.e.
$$
\begin{array}{l}
(-2) \times\ \ 4 \\
+(-3) \times (-17) \\
\underline{+\ \ 2\ \times\ -5} \\
\text{Total} = \text{x.D} = 33
\end{array}
$$

Similarly, –y.D is given by covering the second column of the original matrix, etc., and z.D by covering the third column of the original matrix.

The written work can be reduced by employing one of the methods for summing products given in Chapter 2. The working then appears as follows:

$$
\begin{array}{l}
-8 \\
-14 \\
13
\end{array}
\left(
\begin{array}{cc|cc}
4 & -2 & 3 & 8 \\
4 & -2 & 3 & 8 \\
2 & -3 & 1 & 1 \\
3 & 2 & -1 & 3
\end{array}
\right)
\begin{array}{l}
-5 \\
-17 \\
4
\end{array}
$$

∴ x.D = 33;   –y.D = -33;   z.D = 66;   D = 33

∴ x = y = 1 and z = 2

In practice there is no need to write the original coefficients down again, and the whole working, including the original equations, appears as follows:

$$
\begin{array}{c|ccc|c}
-8 & 4x & -2y & +3z = 8 & -5 \\
-14 & 2x & -3y & +z\ \ = 1 & -17 \\
13 & 3x & +2y & -z\ \ = 3 & 4
\end{array}
$$

∴ D = 33; x.D = 33; –y.D = –33;   z.D = 66

Whence x = y = 1 and z = 2

The next example draws on the well-known rule, that a determinant is unchanged on adding or subtracting rows or columns.

Example 3   Solve
$$2w + 3x + y + 4z = 11$$
$$3w + 5x + 2y + 6z = 19$$
$$2w \quad + y + 4z = 5$$
$$5w + 3x + 3y + 2z = 20$$

We start with: $\det A = D$
$$\begin{vmatrix} w & x & y & z & -1 \\ 2 & 3 & 1 & 4 & 11 \\ 3 & 5 & 2 & 6 & 19 \\ 2 & 0 & 1 & 4 & 5 \\ 5 & 3 & 3 & 2 & 20 \end{vmatrix}$$

The result of adding one row to another (or of subtraction) can be shown by rossing out the current figure and placing the new one alongside.

Successive replacements of the original figure can be allowed to wind around he latter in some consistent pattern. E.g. if the 19 of R_3, C_5 (i.e. Row 3, column 5) is reduced by subtracting 11 from R_2 and then by subtracting 6 from C_4, we could write:

$$\begin{array}{cc} & 2 \\ 8 & 19 \end{array}$$

The working can be written down as follows:

The first six steps of one possible sequence are:

w	x	y	z	−1
2̶ 2 1	−2̶ 3 −2	1̶ 1 1	−4̶ 4 −14	1̶ 11 0
0̶ 3 −4	5̶ 5 5	0̶ 2 -2	8 6 0	10̶ 19 0
2 1	0 0	1 0	4 10	5 1
1̶ 3 5 −6	0̶ 0 3 6	1̶ 2 3 −2	−6̶ −2 2 −4	4 9 20 0

1) $R_5 - R_2$ 2) $R_3 - R_5$ 3) $R_2 - R_3$
4) $R_5 - R_4$ 5) $R_3 - 2R_4$ 6) $R_4 - R_5$

We are now in a position to eliminate all but one figure from C_5; hence the next two steps:

7) $R_5 - R_4 - 3R_2$ 8) $R_2 - R_4$

And now, adding R_4 to R_1 (step 9), we can effectively expand by the last column simply by crossing out C_5 and R_4 (step 10). The minus sign which comes outside the determinant can be ignored, since it does not affect what we are interested in.

This leaves the following situation:

$$\begin{vmatrix} w+1 & x & y & z+1 \\ 1 & -2 & 1 & -14 \\ -4 & 5 & -2 & 0 \\ -6 & 6 & -2 & -4 \end{vmatrix}$$

Three zeros can now be brought into the new column 3 as follows, where the above are now relabelled 1–4, and the columns also 1–4:

Steps
11) $R_4 - R_3$
12) $R_3 + 2R_2$
13) $R_1 - y \times R_2$
14) Cross out R_2, C_3 (i.e. expand by column 3)

The point which has now been reached is:

$$D\begin{vmatrix} w+1 & x+2y & z+10 \\ -y & & +14y \\ -2 & 1 & -28 \\ -2 & 1 & 4 \end{vmatrix}$$

The next step shows that 'z' is zero, since it disappears from the calculation.

Steps
15) $C_1 + 2C_2$

Now expanding by the first column, we have for the determinant:

$D(w + 1 + 3y + 2x)(32)$

Ignoring the 32, and reading off the coefficients of the various terms, since those give their respective values, we have:

$\underline{D = 1; \quad w = 1; \quad x = 2; \quad y = 3; \quad z = 0}$

Substituting in the original equations, these values are found to be correct.

Alternatively, this example could be commenced by a pivotal reduction, using any element in the last column as a pivot—although the '−1' is obviously advantageous. Thereafter the method shown in Example 2 could be applied.

When the figures are simple the method shown in Example 3 can be very effective. That chess can be played blindfolded shows the possibility of such calculations being performed mentally.

ELIMINATION OF ONE VARIABLE

Determinants arise on eliminating unknowns from simultaneous equations- second order determinants on eliminating one unknown, third order on eliminating two unknowns, etc.

We start with the elimination of one variable.

Example 4
$$x + 2y + 3z = 10 \quad\quad\quad …(1)$$
$$2x + y + 4z = 12 \quad\quad\quad …(2)$$
$$3x + 4y + 2z = 19 \quad\quad\quad …(3)$$

a) Eliminating x from Equations (1) and (2) we have:

$$\begin{vmatrix} 1 & 2 \\ 2 & 1 \end{vmatrix} y + \begin{vmatrix} 1 & 3 \\ 2 & 4 \end{vmatrix} z = \begin{vmatrix} 1 & 10 \\ 2 & 12 \end{vmatrix} \quad\quad\quad …(4)$$

Note the connection between the patterns of elements in Equation (4) and in Equations (1) and (2). The first column of each determinant in Equation (4) is the same, and stems from the coefficients of x in Equations (1) and (2). The second column in each case contains the coefficients of the unknown being considered. Registering this pattern, we can immediately eliminate an unknown from any two equations. The next two examples draw on Equations (1), (2) and (3) to demonstrate this.

b) Eliminating x from Equations (2) and (3) we have:

$$5y - 8z = 2$$

c) Eliminating y from Equations (1) and (2) we have:

$$-\begin{vmatrix} 1 & 2 \\ 2 & 1 \end{vmatrix} x + \begin{vmatrix} 2 & 3 \\ 1 & 4 \end{vmatrix} z = \begin{vmatrix} 2 & 10 \\ 1 & 12 \end{vmatrix}$$

The sequence of coefficients being kept the same as in Equations (1) and (2), a minus sign arises with the coefficient of x, because its coefficients are to the left of those for y, in Equations (1) and (2). This amounts to an interchange of columns, which produces the sign change. ∴ $3x + 5z = 14$.

Example 5 $2x + 7y = 39$
             $3x + 8y = 46$

Eliminating x:

$$\begin{vmatrix} 2 & 7 \\ 3 & 8 \end{vmatrix} y = \begin{vmatrix} 2 & 39 \\ 3 & 46 \end{vmatrix}$$

i.e. $\underline{y = \frac{-25}{-5} = 5}$

Eliminating y:

$$-\begin{vmatrix} 2 & 7 \\ 3 & 8 \end{vmatrix} x = \begin{vmatrix} 7 & 39 \\ 8 & 46 \end{vmatrix}$$

∴ $\underline{x = 2}$

PROCEDURE FOR ELIMINATING TWO VARIABLES AT A TIME

Example 6   Eliminate w and x from the following equations:

$$\begin{array}{llll} 4w & + 2x & + 5y & + 3z = 14 \end{array} \qquad \dots(5)$$
$$\begin{array}{llll} 3w & + 7x & + 8y & + 2z = 20 \end{array} \qquad \dots(6)$$
$$\begin{array}{llll} 5w & + 3x & + 9y & + 5z = 22 \end{array} \qquad \dots(7)$$
$$\begin{array}{llll} 7w & + 6x & + 7y & + 4z = 24 \end{array} \qquad \dots(8)$$

To eliminate w and x we can use third order determinants in which the first two columns are recognisable as the coefficients of w and x respectively. The third column consists of the coefficients of the associated unknown. We have, on using Equations (5), (6) and (7):

$$\begin{vmatrix} 4 & 2 & 5 \\ 3 & 7 & 8 \\ 5 & 3 & 9 \end{vmatrix} y + \begin{vmatrix} 4 & 2 & 3 \\ 3 & 7 & 2 \\ 5 & 3 & 5 \end{vmatrix} z = \begin{vmatrix} 4 & 2 & 14 \\ 3 & 7 & 20 \\ 5 & 3 & 22 \end{vmatrix} \qquad \dots(9)$$

And on using Equations (5), (6) and (8) to eliminate w and x, we have:

$$\begin{vmatrix} 4 & 2 & 5 \\ 3 & 7 & 8 \\ 7 & 6 & 7 \end{vmatrix} y + \begin{vmatrix} 4 & 2 & 3 \\ 3 & 7 & 2 \\ 7 & 6 & 4 \end{vmatrix} z = \begin{vmatrix} 4 & 2 & 14 \\ 3 & 7 & 20 \\ 7 & 6 & 24 \end{vmatrix} \qquad \dots(10)$$

Equations (9) and (10) differ only in the third row of each determinant—a point we can use to reduce the amount of calculation.

E.g.
$$\begin{array}{ccc} 22 & -17 & -19 \\ \end{array}$$
$$\begin{vmatrix} 4 & 2 & 5 \\ 3 & 7 & 8 \\ 5 & 3 & 9 \end{vmatrix} = 52$$

And
$$\begin{array}{ccc} 22 & -17 & -19 \\ \end{array}$$
$$\begin{vmatrix} 4 & 2 & 5 \\ 3 & 7 & 8 \\ 7 & 6 & 7 \end{vmatrix} = -81$$

Proceeding to obtain the two equations in two unknowns, we have:

$$52y + 28z = 80$$
$$-81y - 25z = -106$$

Solving, $\underline{y = z = 1}$

Note that the upper left-hand array of coefficients, $\begin{pmatrix} 4 & 2 \\ 3 & 7 \end{pmatrix}$, appeared throughout. In effect this array acted as a (second order) pivot.

The pivotal elements may be chosen from any (rectangular) grouping of four elements from the left-hand side of the equations. The procedure can be illustrated by rearranging the sequence of columns in this last example, as follows:

Example 7   Eliminate w and x from the following equations:

$$5w + 4x + 3y + 2z = 14 \qquad \qquad ...(11)$$
$$8w + 3x + 2y + 7z = 20 \qquad \qquad ...(12)$$
$$9w + 5x + 5y + 3z = 22 \qquad \qquad ...(13)$$
$$7w + 7x + 4y + 6z = 24 \qquad \qquad ...(14)$$

Eliminating w and x from Equations (11), (12) and (13), we have:

$$\begin{vmatrix} 5 & 4 & 2 \\ 8 & 3 & 7 \\ 9 & 5 & 3 \end{vmatrix} y - \begin{vmatrix} 4 & 3 & 2 \\ 3 & 2 & 7 \\ 5 & 5 & 3 \end{vmatrix} z = \begin{vmatrix} 4 & 2 & 14 \\ 3 & 7 & 20 \\ 5 & 3 & 22 \end{vmatrix}$$

The sign associated with each determinant can be obtained by considering the positioning of the 'pivotal columns'- those columns which contain the pivotal elements, $\begin{pmatrix} 4 & 2 \\ 3 & 7 \end{pmatrix}$. For each determinant we ask the question, "how many column changes are needed to bring the 'pivotal columns' into the first two columns?"

For $\begin{vmatrix} 5 & 4 & 2 \\ 8 & 3 & 7 \\ 9 & 5 & 3 \end{vmatrix}$ two changes of column suffice, therefore its sign is +.

For $\begin{vmatrix} 4 & 3 & 2 \\ 3 & 2 & 7 \\ 5 & 5 & 3 \end{vmatrix}$ one change of column suffices, therefore its sign is minus.

For $\begin{vmatrix} 4 & 2 & 14 \\ 3 & 7 & 20 \\ 5 & 3 & 22 \end{vmatrix}$ no column changes are required, therefore its sign is +.

In general, the sign of a determinant is given by minus one raised to the power of the number of column changes required to bring the pivotal columns into the leading places.

This method of using a second order determinant as pivot is especially attractive if it contains a zero (or better still two, in opposite corners), especially if the other diagonal contains simple multiples of unity—the simpler the better. This desirable feature can be brought about by row or column operations. (But note that a pivot with zeros throughout one row or column cannot be used, since its determinant is zero, and division by zero is disallowed.) Column operations correspond to a change of variable. E.g. if, in the last example, we start by taking column 2 from column 1, then the unknown in column 1 becomes $y - w$. The 'higher order determinant' method makes this clear, and is simple to use if column operations are contemplated.

Exercise
Solve the following:

1) $4x + 8y = 20$
    $2x + 5y = 12$ [x=1, y=2]

2) $4x + 2y = 14$
    $6x + 9y = 27$ [x=3, y=1]

3) $2x + 9y = 36$
    $9x + 3y = 12$ [x=0, y=4]

4) $3x + 2y + z = 10$
    $5x + 3y + 2z = 17$
    $7x + 8y + z = 26$ [x=1, y=2, z=3]

5) $4x + 2y + 2z = 22$
    $2x + 4y + 7z = 38$
    $8x + 4y + 2z = 36$ [x=3, y=1, z=4]

6) $4x + 2y + 6z = -8$
    $2x + 7y + 3z = 8$
    $4x + 8y + 2z = 20$ [x=3, y=2, z= -4]

7) $2x - 4y + 6z = -8$
    $4x - 2y - 5z = 13$
    $-2x + 4y + 6z = -4$ [x=3, y=2, z= -1]

8) $2x + y + 3z + 5w = 76$
    $3x + 2y - 4z + 3w = 5$
    $2x + 4y - 6z + 2w = -26$
    $8x - 2y + 7z - w = 81$ [x=4, y= -1, z=8, w=9]

9) $x + y + 3z + 4w = 26$
    $2x + 4y + 6z + 8w = 58$
    $x + 3y + 5z + 7w = 48$
    $4x + 3y + 2z + w = 21$ [x=1, y=3, z=2, w=4]

10) $2x + 3y + 8z + 9w = -4$
    $4x + y + 2z + 9w = -24$
    $2x + 3y + 2z + 3w = 2$
    $4x + 9y + 3z + 4w = 15$ [x=1, y=2, z=3, w= -4]

Chapter 5

INVERSION OF MATRICES

It is well-known that the inverse of a 2 × 2 matrix can be written straight down. The same procedure can be extended to allow inverses of third and fourth degree square matrices to be written straight down. Use of this is made in the following examples.

Example 1   Given $A = \begin{pmatrix} 1 & 3 \\ 2 & 4 \end{pmatrix}$

$\square$ $A^{-1} = \frac{1}{1 \times 4 - 2 \times 3} \begin{pmatrix} 4 & -3 \\ -2 & 1 \end{pmatrix}$

The following explanation of the steps applies also to higher order matrices.

STEPS FOR DETERMINING THE ELEMENTS OF A^{-1} FROM THE ELEMENTS OF

1) Along the leading diagonal (from top left to bottom right) replace each element by its cofactor. (The reader is reminded that the cofactor of the element in the ith row and the column is obtained by crossing out the row and column containing the element, taking the determinant of what remains, and multiplying by $(-1)^{i+j}$).

2) Elsewhere replace each element by the cofactor of its mirror-image in the leading diagonal. E.g. the mirror-image of 2 is 3 in the matrix above, and the cofactor of this 3 is –2.

3) Divide by the determinant of A, $|A|$.

Example 2 Find the inverse of $B = \begin{pmatrix} 1 & 2 & 3 \\ 4 & 5 & 2 \\ 3 & 1 & 4 \end{pmatrix}$

Solution: $B^{-1} = \frac{-1}{35} \begin{pmatrix} 18 & -5 & -11 \\ -10 & -5 & 10 \\ -11 & 5 & -3 \end{pmatrix}$

STEPS

1) Replace the top left-hand digit by its cofactor: $\begin{vmatrix} 5 & 2 \\ 1 & 4 \end{vmatrix} = 18$.

Similarly, for other elements on the leading diagonal:

replace 5 by $\begin{vmatrix} 1 & 3 \\ 3 & 4 \end{vmatrix} = -5$, and 4 by $\begin{vmatrix} 1 & 2 \\ 4 & 5 \end{vmatrix} = -3$.

2) Other elements are replaced by the cofactors of their mirror-images in the leading diagonal. E.g. in the first column, the 4 is replaced by the cofactor of 2, viz $-\begin{vmatrix} 4 & 2 \\ 3 & 4 \end{vmatrix} = -10$, and the 3 is replaced by the cofactor of the top right-hand 3, viz $\begin{vmatrix} 4 & 5 \\ 3 & 1 \end{vmatrix} = -11$.

3) The divisor, $|B|$, is obtained from the first row of B times the first column of B^{-1}. More generally, the ith row of B times the ith column of B, or vice-versa, yields $|B|$.

CHECKS

These are provided by the considerations that the ith row of b times the jth column of b, or vice-versa:

 (i) comes to zero when $i \neq j$

 (ii) $= |B|$ when $i = j$

Example 3 Invert $C = \begin{pmatrix} 1 & 0 & 2 & 3 \\ 2 & 5 & 1 & 4 \\ 3 & 1 & 0 & 2 \\ 1 & 3 & 1 & 4 \end{pmatrix}$

Solution: $C = 3 \begin{matrix} 5 \\ 1 \\ 1 \\ 8 \end{matrix} \begin{matrix} -13 \end{matrix} \begin{pmatrix} 1 & 0 & 2 & 3 \\ 2 & 5 & 1 & 4 \\ 3 & 1 & 0 & 2 \\ 1 & 3 & 1 & 4 \end{pmatrix} \begin{matrix} 5 \\ 2 \\ -2 \end{matrix} \begin{matrix} 4 \\ 0 \end{matrix} 5$

$\square\ C^{-1} = \dfrac{-1}{33} \begin{pmatrix} -4 & -7 & -10 & 15 \\ 2 & -13 & 5 & 9 \\ -22 & -22 & 11 & 33 \\ 5 & 17 & -4 & -27 \end{pmatrix}$

Here cofactors are evaluated using the pyramidal method of evaluating determinants.

WHY THE METHOD WORKS

If we multiply the first row of C by the first column of C^{-1}, we are multiplying each element of C by its cofactor, and summing, which should give us |C| (expansion of the determinant by first row). Similarly the second row of C times the second column of C^{-1} should give us |C|, and likewise the third row of C times the third column of C^{-1}, etc.

The above mentioned multiplications give: $CC^{-1} = \dfrac{1}{|C|} \begin{pmatrix} |C| & . & . & . \\ . & |C| & . & . \\ . & . & |C| & . \\ . & . & . & |C| \end{pmatrix} = \begin{pmatrix} 1 & . & . & . \\ . & 1 & . & . \\ . & . & 1 & . \\ . & . & . & 1 \end{pmatrix}$

Off the leading diagonal we find that we are multiplying the elements of one row of C by the cofactors of another row (and summing). This is equivalent to evaluating a determinant with two rows the same—which comes to zero. I.e. all other elements- represented by dots—are zero. E.g. the second column of C^{-1} consists of the cofactors of the second row of C, which uses the elements,

$$\begin{pmatrix} 1 & 0 & 2 & 3 \\ . & . & . & . \\ 3 & 1 & 0 & 2 \\ 1 & 3 & 1 & 4 \end{pmatrix}$$

If the first row of C is multiplied by the second column of C^{-1} the row (1 0 2 3)   is   used twice, and we are evaluating the determinant,

$$\begin{vmatrix} 1 & 0 & 2 & 3 \\ 1 & 0 & 2 & 3 \\ 3 & 1 & 0 & 2 \\ 1 & 3 & 1 & 4 \end{vmatrix} = 0$$

With 5×5 matrix inversion there is more difficulty because each cofactor is a 4×4 determinant. On inverting a matrix, D, e.g., the cofactors of the last column of D^{-1} draw on elements from the first four rows of D. These can be horizontally partitioned in the centre, and values of 2×2 determinants placed pyramidally above and below, as shown in Chapter 3. Proceeding thus, we shall be dealing with five groups of four rows, requiring a total of five 'pyramids', some above and some below.

This process is somewhat cumbersome for degree 5, and even more so for higher degree matrices.

Exercise Invert the following matrices:—

1) $M = \begin{pmatrix} 1 & 2 \\ 3 & 4 \end{pmatrix}$ $\left[M^{-1} = \frac{1}{-2}\begin{pmatrix} 4 & -2 \\ -3 & 1 \end{pmatrix} \right]$

2) $M = \begin{pmatrix} 5 & 2 \\ 3 & 4 \end{pmatrix}$ $\left[M^{-1} = \frac{1}{14}\begin{pmatrix} 4 & -2 \\ -3 & 5 \end{pmatrix} \right]$

3) $M = \begin{pmatrix} 2 & 1 & 3 \\ 4 & 5 & 2 \\ 1 & 2 & 4 \end{pmatrix}$ $\left[M^{-1} = \frac{1}{27}\begin{pmatrix} 16 & 2 & -13 \\ -14 & 5 & 8 \\ 3 & -3 & 6 \end{pmatrix} \right]$

4) $M = \begin{pmatrix} 3 & 2 & 1 \\ 0 & 4 & 2 \\ 1 & 7 & 2 \end{pmatrix}$ $\left[M^{-1} = \frac{-1}{18}\begin{pmatrix} -6 & 3 & 0 \\ 2 & 5 & -6 \\ -4 & -19 & 12 \end{pmatrix} \right]$

5) $M = \begin{pmatrix} 4 & 3 & 2 \\ 1 & 4 & 2 \\ 3 & 2 & 1 \end{pmatrix}$ $\left[M^{-1} = \frac{-1}{5}\begin{pmatrix} 0 & 1 & -2 \\ 5 & -2 & -6 \\ -10 & 1 & 13 \end{pmatrix} \right]$

6) $M = \begin{pmatrix} 7 & 11 & 9 \\ 2 & 8 & 4 \\ 4 & 9 & 8 \end{pmatrix}$ $\left[M^{-1} = \frac{1}{70}\begin{pmatrix} 28 & -7 & -28 \\ 0 & 20 & -10 \\ -14 & -19 & 34 \end{pmatrix} \right]$

7) $M = \begin{pmatrix} 3 & 4 & 2 & 1 \\ 4 & 1 & 2 & 3 \\ 4 & 3 & 1 & 2 \\ 3 & 2 & 1 & 4 \end{pmatrix}$ $\left[M^{-1} = \frac{1}{40}\begin{pmatrix} -11 & 7 & 21 & -13 \\ 9 & -13 & 1 & 7 \\ 19 & 17 & -29 & -3 \\ -1 & -3 & -9 & 17 \end{pmatrix} \right]$

8) $M = \begin{pmatrix} 4 & 9 & 2 & 5 \\ 3 & 2 & 2 & 1 \\ 4 & 6 & 7 & 2 \\ 8 & 9 & 1 & 1 \end{pmatrix}$ $\left[M^{-1} = \frac{1}{432}\begin{pmatrix} -23 & 216 & -57 & 13 \\ 13 & -216 & 51 & 49 \\ -24 & 0 & 62 & -24 \\ 91 & 216 & -75 & -89 \end{pmatrix} \right]$

9) $M = \begin{pmatrix} 5 & 2 & 1 & 0 \\ 3 & 8 & 7 & 1 \\ 2 & 4 & 3 & 3 \\ 4 & 2 & 9 & 3 \end{pmatrix}$ $\left[M^{-1} = \frac{-1}{714}\begin{pmatrix} -156 & -42 & 2 & -16 \\ 6 & -84 & -55 & 83 \\ 54 & -42 & 100 & -86 \\ 42 & 126 & -266 & -14 \end{pmatrix} \right]$

10) $M = \begin{pmatrix} 4 & 2 & 1 & 8 \\ 7 & 2 & 3 & 1 \\ 2 & 6 & 5 & 0 \\ 4 & 8 & 2 & 1 \end{pmatrix}$ $\left[M^{-1} = \frac{1}{1364}\begin{pmatrix} -36 & 220 & -164 & 68 \\ -13 & -110 & -17 & 214 \\ 30 & 44 & 354 & 284 \\ -188 & -88 & 38 & -52 \end{pmatrix} \right]$

Chapter 6

CURVE-FITTING

In this chapter we briefly consider both mathematical and statistical curve-fitting. The former deals with exact fit, and the latter with cases where exact fit is not possible. As is customary the least squares criterion is employed in the latter case.

The material of this chapter provides an illustration of points already made concerning the VERTICALLY AND CROSSWISE formula. There are interesting methods of precise curve-fitting which come under other sutras but this is not the place to consider them.

(a) MATHEMATICAL OR EXACT CURVE-FITTING

Example 1   Find a quadratic which passes through the three points: (8,7)
(6,5)
(9,4)

> **Method (i)**   Changing the origin to (6,5), the three points become: (2,2)
> (0,0)
> (3,–1)
>
> Let the equation of the line through these points be:
> $Y = AX^2 + BX$,
> and then we have: $2 = 4A + 2B$
> $-1 = 9A + 3B$
>
> Hence $6y = -8X^2 + 22X$ ...(1)
> This is the required equation, using the new origin.
>
> Or, more fully: $-\begin{vmatrix} 4 & 2 \\ 9 & 3 \end{vmatrix} Y = \begin{vmatrix} 2 & 2 \\ 3 & -1 \end{vmatrix} X^2 + (-) \begin{vmatrix} 4 & 2 \\ 9 & -1 \end{vmatrix} X$

Equation (1) may meet our needs, or we may prefer to use the original variables, x and y, when we have:

$6(y-5) = -8(x-6)^2 + 22(x-6)$

i.e. $\underline{6y = -8x^2 + 118x - 390}$

And this is the required equation, in terms of the original co-ordinates.

Method (ii)   Inserting the values for the three points into the equation: $ax^2 + bx + c = y$, we have

96	$8\%8a + 8b + 1 = 7$	-2
72	$6\%6a + 6b + 1 = 5$	3
-162	$9\%9a + 9b + 1 = 4$	-1

Here the higher order determinant procedure is being employed for solving the equations, one row being understood but not written.

Solving, we immediately write: $6y = -8x^2 + 118x - 390$, which is the required equation.

(b) REGRESSION

Least Squares Linear Regression of y on x

Given n points, (x_1,y_1), (x_2,y_2), ... , (x_i,y_i), ... , (x_n,y_n), it is desired to fit a straight line, $y = a + bx$, using the 'least squares' criterion. I.e. we minimise:

$$S^2 = \sum (y_i - a - bx_i)^2 \qquad \qquad ...(1)$$

The formulae can be written straight down, starting with the regression line equation:

$$y - a - bx = 0 \qquad \qquad ...(2)$$

Normal $\left\{ \begin{array}{l} \sum y - na - b\sum x = 0 \\ \sum xy - a\sum x - b\sum x^2 = 0 \end{array} \right\}$ ...(3)
Equations   ...(4)

Formula (3) is obtained from Equation (2) by summation, and
Formula (4) is obtained by 'multiplying' Formula (3) by x, noting that $n = \sum f_i$ where f denotes frequency, and $f_i = 1$ in this case, for all i from 1 to n.

Justification for Obtaining the Formulae in this Way

Since S^2 is to be minimised, and there are two variables for this purpose, 'a' and 'b', we note that:

(i) $\frac{\partial}{\partial a}(y - a - bx) = -1,$

(ii) $\frac{\partial}{\partial b}(y - a - bx) = -x,$

Hence, putting $\frac{\partial(S^2)}{\partial a} = \frac{\partial(S^2)}{\partial b} = 0$, we have:

$$\frac{\partial(S^2)}{\partial a} = \sum(y - a - bx) = 0 \text{ and } \frac{\partial(S^2)}{\partial b} = \sum x(y - a - bx) = 0$$

The minus signs being irrelevant, Equations (3) and (4) follow.

Example 2   Fit a regression line of y on x to the four pairs of points given below. The complete working is written down as follows:

$$
\begin{array}{cc}
y & x \\
8 & 5 \\
7 & 6 \\
4 & 7 \\
3 & 9 \\
\end{array}
$$

$$\left\{ \begin{array}{l} 22 = 4a + 27b \\ 137 = 27a + 191b \end{array} \right\} \quad \begin{array}{l} \text{Normal} \\ \text{equations} \end{array}$$

Solving these simultaneously,

$$-135y = -3x + 46 \quad \text{(Regression line of y on x)}$$

The steps are as follows, using the procedures of Chapter 2 to evaluate sums of squares and sums of products.
Under the y-column we write $\sum y$ (=22), $\sum yx$ (=137).
Under the x-column we write $\sum x$ (=27) and $\sum x^2$ (=191).
And between these two totals we put n (=4) and $\sum x$ (=27).
Finally we solve simultaneously for a and b.

Note that for the least squares linear regression of x on y the roles of x and y are interchanged, and we have:

Regression line of x on y: $x - a - by = 0$

$$\begin{array}{l} \text{Normal} \\ \text{equations} \end{array} \left\{ \begin{array}{l} \sum x - na^\Re - b^\Re \sum y = 0 \\ \sum xy - a^\Re \sum y - b^\Re \sum y^2 = 0 \end{array} \right\}$$

Least Squares Quadratic Regression of y on x

This procedure differs from the linear case in that there is an extra term, c, the coefficient of x^2.

Hence an extra equation arises from $\frac{\partial(s^2)}{\partial c}$ which is determined by $\frac{\partial}{\partial c}(y - a - bx - cx^2) = -x^2$. The upshot is that the third 'Normal equation' is obtained from the second by 'multiplying' once again by x.

So we have:

$$\text{Regression line:} \quad y - a - bx - cx^2 = 0$$

Normal equations for a quadratic
$$\begin{cases} \Sigma y - na - b\Sigma x - c\Sigma x^2 = 0 \\ \Sigma xy - a\Sigma x - b\Sigma x^2 - c\Sigma x^3 = 0 \\ \Sigma x^2 y - a\Sigma x^2 - b\Sigma x^3 - c\Sigma x^4 = 0 \end{cases}$$

In practical examples, the coefficients of the Normal equations can be calculated and written down below the given pairs of values of x and y—as was done in the linear case. Under the y-column we write, respectively, the values of $\sum y$, $\sum xy$ and $\sum x^2 y$. Under the x-column we write, respectively, $\sum x$, $\sum x^2$ and $\sum x^3$.. And under the x^2 column we write, respectively, $\sum x^2$, $\sum x^3$ and $\sum x^4$.

Example 3   Fitting a quadratic equation to the four points of Example 2, the complete working can be written down as follows:

y	x	x^2
8	5	25
7	6	36
4	7	49
3	9	81

$$22 = 4a + 27b + 191c$$
$$137 = 27a + 191b + 1{,}413c$$
$$891 = 191a + 1{,}413b + 10{,}883c$$

Solving, the quadratic regression line of y on x is:

$$\underline{11{,}880y = 312{,}444 - 57{,}618x + 2{,}970x^2}$$

Regression Line of Third Degree

The same considerations as before when applied to this case yield the following equations:

$$y - a - bx - cx^2 - dx^3 = 0 \qquad \text{(Regression line of y on x)}$$

$$\left.\begin{array}{l} \Sigma y - na - b\Sigma x - c\Sigma x^2 - d\Sigma x^3 = 0 \\ \Sigma xy - a\Sigma x - b\Sigma x^2 - c\Sigma x^3 - d\Sigma x^4 = 0 \\ \Sigma x^2 y - a\Sigma x^2 - b\Sigma x^3 - c\Sigma x^4 - d\Sigma x^5 = 0 \\ \Sigma x^3 y - a\Sigma x^3 - b\Sigma x^4 - c\Sigma x^5 - d\Sigma x^6 = 0 \end{array}\right\} \text{'Normal equations' for a cubic}$$

In arranging calculations, the pattern of columns starts the same as before, differing only in the last column: y, space, x, x^2, space, x^4.

The same principles apply to higher degree regression lines.

Exercises

Mathematical or Exact Curve-Fitting.

Find the quadratics through the following points:

1) (4,3),   (6,4),   (8,9) $[2y = x^2 - 9x + 26]$
2) (8,2),   (3,2),   (5,6) $[3y = -2x^2 + 22x - 42]$
3) (1,2),   (3,4),   (5,6) $[y = x + 1]$
4) (3,7),   (2,4),   (4,6) $[y = -2x^2 + 13x - 14]$
5) (4,4),   (2,2),   (3,9) $[y = -6x^2 + 37x - 48]$
6) (7,4),   (2,4),   (8,3) $[-6y = x^2 - 9x - 10]$
7) (10,12), (11,3),   (6,9) $[20y = -39x^2 + 639x - 2250]$
8) (2,3),   (9,6),   (4,5) $[35y = -4x^2 + 59x + 3]$
9) (3,7),   (8,4),   (2,5) $[30y = -13x^2 + 115x - 48]$
10) (11,2), (3,7),   (9,3) $[48y = 2x^2 - 44x + 459]$

Regression.

Fit a regression line of y on x to each of the following sets of points:

11) (2,4),   (6,8),   (8,3),   (9,2) $[115y = -33x + 695]$
12) (10,12), (3,6),   (2,9),   (7,3) $[41y = 21x - 192]$
13) (4,1),   (2,6),   (7,8),   (9,4) $[58y = 7x + 237]$
14) (3,2),   (1,4),   (5,2),   (3,7) $[4y = -2x + 21]$

Chapter 7

EVALUATION OF LOGARITHMS AND EXPONENTIALS

We consider natural logarithm (denoted by 'ln'), and start by demonstrating a method for evaluating ln1.2 (=0.18232), and its inverse, exp(0.18232), and afterwards proceed to a fuller account of the systems proposed for evaluating logarithms and exponentials generally.

Example 1   Evaluate ln1.2

The complete working is as follows

$$\overset{0}{}\quad\overset{2}{}\quad\overset{-2}{}\quad\overset{\frac{8}{3}}{}\quad\overset{-4}{}\quad\overset{6.4}{}$$

Let $\ln(1+2x) = a + bx + cx^2 + dx^3 + ex^4 + fx^5 + \ldots$ 　　　　　...(1)

Differentiating with respect to x:

$$\overset{2}{}\quad\overset{-4}{}\quad\overset{8}{}\quad\overset{-16}{}\quad\overset{32}{}\quad\overset{-64}{}$$

$\frac{2}{1+2x} = b + 2cx + 3dx^2 + 4ex^3 + 5fx^4 + 6gx^5 + \ldots$ 　　　　　...(2)

$\therefore$ $\underline{\ln1.2 = 0.18232}$

Explanation
The justification for Equation (1) is Taylor's theorem.

Putting x = 0.1 gives ln1.2 = a.bcd...., where the figures customary in decimal notation have been replaced by letters for generality.
On putting x=0 in Equation (1) we obtain a=0, and the zero is written above the 'a'.

We now successively equate coefficients of 1, x, x², . . . in Equation (2), applying mental VERTICALLY AND CROSSWISE multiplication by 1+2x on the right-hand side. In this way we successively obtain the coefficients:

$b = 2$ from the coefficient of unity,
$2c = -4$ from the coefficient of x,
$3d = 8$ from the coefficient of x^2, etc.

Finally, putting $x = 0.1$ in Equation (1), the required value follows, namely, ln1.2 = 0.18232.

A check is obtained by calculating exp(0.18232). The method is similar to the foregoing.

Example 2   Evaluate exp(0.18232)

To keep the figures small, we can use $0.18232 = 0.2\bar{2}232$
The complete working is as follows.
Let $\ln(A + Bx + Cx^2 + Dx^3 + \ldots) = 0 + 2x + \bar{2}x^2 + 2x^3 + 3x^4 + 2x^5$ ...(3)

Differentiate with respect to x:

$$\begin{array}{cccccc} 2 & 0 & \bar{2} & 22\tfrac{2}{3} & 48 \end{array}$$
$$B+2Cx+3Dx^2+4Ex^3+5Fx^4+6Gx^5+\ldots$$
$$\therefore \quad A+Bx+Cx^2+Dx^3+Ex^4+Fx^5+ \quad = 2 - 4x + 6x^2 + 12x^3 + 10x^4 \quad ...(4)$$
$$\begin{array}{cccccc} 1 & 2 & 0 & -\tfrac{2}{3} & 5\tfrac{2}{3} & 9.6 \end{array}$$

☐ exp(0.18232) ≐ 1.2000

Explanation
Putting x=0 in Equation (3) gives A = 1. This is placed below the A in Equation (4). Now successively equate coefficients of 1, x, x^2 etc. (as before), to obtain:

$B = 2$ from the coefficient of unity—written above the B and below the Bx on the left-hand side,

$2C = 0$ from the coefficient of x– and '0' is written above the 2C and below the C on the left-hand side,

$3D = \bar{2}$ from the coefficient of x^2– and $\bar{2}$ is written above the 3D and below the D on the left-hand side,
                and so on.

These examples illustrate methods which are well-suited to the determination of logs of numbers close to unity, or of powers of 'e' close to zero. We need to broaden the system, and what follows shows how this can be done.

For further calculation we can usefully draw on a table consisting of the natural logarithms of 2, 3, 4, ...10.

How this table can be built up will be shown later. First we consider its use.

Number	Natural Logarithm
2	0.693 147 180
3	1.098 612 289
4	1.386 294 361
5	1.609 437 912
6	1.791 759 469
7	1.945 910 149
8	2.079 441 542
9	2.197 224 577
10	2.302 585 509

Example 3   Evaluate ln 63

Since $63 = 9 \times 7$, $\ln 63 = \ln 9 + \ln 7 = \underline{4.143\ 134\ 726}$

Example 4   Evaluate ln 0.05

Since $0.5 = \frac{5}{9}$, $\ln 0.05 = \ln 5 - \ln 9 - \ln 10 = \underline{-2.890\ 372\ 174}$

These examples can be worked out to a degree of precision limited only by the precision of the above table. Some of the examples which follow are a little less obviously obtainable from the table.

Example 5   Evaluate ln 864

Since 8 and 64 are both divisible by 8, we can say,

$$864 = 8 \times 108 = 8 \times 9 \times 12 = 2^5 \times 3^3$$

$\therefore$ $\ln 864 = 5\ln 2 + 3\ln 3 = 5 \times 0.693\ 147 + 3 \times 1.098\ 612 = \underline{6.761\ 57}$

Example 6   Evaluate ln 375

Here the 75 is a clue: $375 = \frac{15}{4} \times 100$

$\therefore$ $\ln 375 = \ln 5 + \ln 3 + 2\ln 10 - \ln 4 = \underline{5.926\ 926}$

Example 7   Evaluate ln 0.714285

A lot of work is saved here by spotting the recurring decimal for $\frac{5}{7}$ (see 'Vedic

Mathematics', Chapter 26).

$$\therefore \ln 0.71428\bar{5} = \ln 5 - \ln 7 = \underline{-0.336\ 472}$$

Now suppose that we wish to evaluate the natural logarithm of a number which is quite close to being obtainable by the above method.

Example 8   Evaluate ln238

Since $8 \times 3 = 24$, consider $\frac{238}{24} = 10.\overline{116}$

$$\therefore \ln 238 = \ln\left(24 \times \tfrac{238}{24}\right) = \ln 24 + \ln 10 + \ln 1.0\overline{123}$$

Note that $0.0016 = 0.00\overline{23}$

$$\text{Let } \ln(1 + 0x - x^2 + 2x^3 - 3x^4 - 3x^5 \ldots) \overset{0\quad 0\quad -1\quad 2\quad -\frac{7}{2}\quad -1}{=} a + bx + cx^2 + dx^3 + ex^4 + fx^5 + \ldots .$$

Differentiating w.r.t. 'x':

$$\frac{-2x+6x^2-12x^3-15x^4\ldots}{1+0x-x^2+2x^3-3x^4-3x^5\ldots} \overset{0\quad -2\quad 6\quad -14\quad -5\quad -22}{=} b + 2cx + 3dx^2 + 4ex^3 + 5fx^4 + 6gx^5 + \ldots$$

And now, $\ln 1.0\overline{123} = 0.0\overline{12}\,\overline{36}$

$$
\begin{array}{ll}
\ln 8 & = 2.079\ 44 \\
\ln 3 & = 1.098\ 61 \\
\ln 10 & = 2.302\ 59 \\
\text{Summing: } \underline{\ln 238} & = \underline{5.472\ 3}
\end{array}
$$

Example 9   Evaluate ln1.913

We can make use of: $2 \times \frac{1.913}{2} = 2 \times 0.9565 = 2 \times 1.0\overline{435}$

This enables us to work with a number close to unity.

$$\text{Let } \ln(1 + 0x - 4x^2 - 3x^3 - 5x^4) \overset{0\quad 0\quad -4\quad -3\quad -13\quad -12\quad -\frac{275}{6}}{=} a + bx + cx^2 + dx^3 + ex^4 + fx^5 + gx^6 + \ldots$$

$$\square \quad \frac{-8x-9x^2-20x^3}{1+0x-4x^2-3x^3-5x^4} \overset{0\quad -8\quad -9\quad -52\quad -60\quad -275}{=} b + 2cx + 3dx^2 + 4ex^3 + 5fx^4 + 6gx^5 + \ldots$$

$$
\begin{array}{ll}
\therefore \ln 0.9565 & = 0.0\overline{44}\ \overline{45} \\
\text{and } \ln 2 & = 0.693\ 15 \\
\text{Adding, } \underline{\ln 1.913} & = \underline{0.648\ 7}
\end{array}
$$

A more accurate answer can be obtained using a closer approximation to 1.913, such as **1.92 = 3 × 0.64**. Indeed, the greater the accuracy sought, the closer we need to make this initial approximation. The factor limiting accuracy then is the number of figures available in the initial table.

To evaluate exponentials which are not close to unity we can draw on the same table.

Example 10   Evaluate A = exp3.421

First we seek that sum or difference of (standard) logs which is as close as possible to 3.421.

Drawing on ln5 and ln6, we have: $\ln 30 = \ln 5 + \ln 6 = 3.401\ 197\ 381$

We wish to solve $\ln A = 3.421$

$\therefore \ln\left(\frac{A}{30}\right) = 0.020\ \overline{2}03\ \overline{4}2\overline{1}$

Let $\ln\left(\frac{A}{30}\right) = \ln(a+bx + cx^2 + dx^3 + ex^4 + fx^5 + gx^6 + hx^7 + ...)$
$= 0+0x+ 2x^2 + 0x^3 - 2x^4 + 0x^5 + 3x^6 - 4x^7 + 2x^8 - x^9$

- Differentiating w.r.t. 'x':

$$0 \quad 4 \quad 0 \quad 0 \quad 0 \quad 2 \quad -28 \quad 53\tfrac{1}{3} \quad -81$$

$$\frac{b+2cx+3dx^2+4ex^3+5fx^4+6gx^5+7hx^6+8ix^7+9jx^8+...}{a+bx+cx^2+dx^3+ex^4+fx^5+gx^6+hx^7+ix^8+jx^9+...} = 0+4x+0x^2-8x^3+0x^4+18x^5-28x^6+16x^7-9x^8$$

$$1 \quad 0 \quad 2 \quad 0 \quad 0 \quad 0 \quad \tfrac{1}{3} \quad -4 \quad 6\tfrac{2}{3} \quad -9$$

$\therefore \exp(3.421) = 30 \times 1.020\ 000\ 0\overline{1}$
$= \underline{30.599\ 999\ 7}$

It is a pleasant surprise to find that the simple table of logs in reverse has advantages of flexibility, which compensate for its indirectness. This is borne out by the next few examples.

Example 11   Evaluate B = exp(23.51)

First note that $10\ln 10 = 23.0$ to one decimal place,
and that $\ln 10 - \ln 6 = 0.5$ to one decimal place.

Hence the following starting point:
$10\ln 10 = 23.025\ 86$
and $\ln 10 - \ln 6 = 0.506\ 82$

Adding: $11\ln 10 - \ln 6 = 23.536\ 68$ and $\ln B = 23.51$ is required.

Subtracting: $\ln\left(\frac{6B}{10^{11}}\right) = -0.026\,68$

$$= -0.03\overline{3}\,\overline{3}2$$

Let $\ln(\frac{6B}{10^{11}}) = \ln(a + bx + cx^2 + dx^3 + ex^4 + fx^5 + ...)$
$$= -3x^2 + 3x^3 + 3x^4 + 2x^5$$

Differentiating w.r.t. 'x':

$$\begin{array}{ccccccc} 0 & -6 & 9 & 30 & -35 & -54 & 124.5 \end{array}$$
$$\frac{b+2cx+3dx^2+4ex^3+5fx^4+6gx^5+7hx^6+...}{a+bx+cx^2+dx^3+ex^4+fx^5+gx^6+hx^7+...} \qquad = -6x + 9x^2 + 12x^3 + 10x^4$$
$$\begin{array}{ccccccc} 1\ 0 & -3 & 3 & 7.5 & -7 & -9 & 17.81 \end{array}$$

$\therefore \frac{6B}{10^{11}} = 1.0\overline{3}3\,7\overline{2}8$

$\therefore \underline{B = \exp(23.51) = 1.6\dot{2}2\,79 \times 10^{10}}$

Example 12 Evaluate exp(−0.888)

We can try using $\ln3 - \ln7 = 0.\overline{9}53\,\overline{3}02\,140$
Required: $\ln A = 0.\overline{9}12$
Subtracting: $\ln\left(\frac{7A}{3}\right) = 0.0\overline{4}1\,30\overline{2}\,\overline{1}40$

Let $\ln(a + bx + cx^2 + ...) = -4x^2 - x^3 + 3x^4 + 0x^5 - 2x^6 - x^7 - 4x^8$

Differentiating w.r.t.'x':

$$\begin{array}{cccccccc} 0 & -8 & -3 & 44 & 20 & -145 & -84 & -61 \end{array}$$
$$\frac{b+2cx+3dx^2+4ex^3+5fx^4+6gx^5+7hx^6+8ix^7...}{a+bx+cx^2+dx^3+ex^4+fx^5+gx^6+hx^7+ix^8...} = 0 - 8x - 3x^2 + 12x^3 + 0x^4 - 12x^5 - 7x^6 - 32x^7$$
$$\begin{array}{cccccccc} 1\ 0 & -4 & -1 & 11 & 4 & -24.16 & 12 & -7.625 \end{array}$$

$\therefore \frac{7A}{3} = 1.0\overline{4}01254\overline{3}$

$\underline{A = 0.411\,477\,7}$

Let us now consider how the natural logs 2, 3, 4, ... 10 can be calculated, using the procedure demonstrated in Example 1.

We have: $2 = \frac{6}{5} \times \frac{5}{4} \times \frac{4}{3}$ ...(5)

$$= 1.2 \times 1.25 \times 1.3$$

Since our method enables us to calculate ln1.2, ln1.25 and ln1.3, (all of these being sufficiently close to unity), we obtain ln2 from Equation (5) above.

Hence also $\ln 4 = 2\ln 2$
and $\ln 8 = 3\ln 2$.

And knowing $\ln\frac{4}{3}$ and $\ln\frac{5}{4}$ we can obtain $\ln 3$ and $\ln 5$.
Also $\ln 9 = 2\ln 3$
and $\ln 10 = \ln 2 + \ln 5$

Finally, noting that $\frac{2\times 7 \times 7}{10\times 10}$ we can obtain $\ln 7$ by evaluating $\ln 0.98$.

Alternative

Instead of $2 = \frac{6}{5}\times\frac{5}{4}\times\frac{4}{3}$, for which the largest ratio is $\frac{4}{3}$, we can use $2 = \frac{10}{9}\times\frac{9}{8}\times\frac{8}{7}\times\frac{7}{6}\times\frac{6}{5}$, where the largest ratio is $\frac{6}{5} = 1.2$. Similarly other sequences of numbers can be employed, such as $\frac{9}{8}\times\frac{8}{7}\times\frac{7}{6} = \frac{3}{2}$, which yields $\ln 3 - \ln 2$, and hence $\ln 3$ once $\ln 2$ has been calculated.

Note that the methods given include the expansion of logs and exponents **of polynomials** as power series. Indeed it is a feature of this system that in learning methods of calculation we are also learning algebraic methods. Chapter 15 of this book gives some applications of this facility for obtaining functions of polynomials as power series: it enables us to obtain power series solutions to non-linear differential equations, and transcendental equations.

Exercises

Evaluate the following to four decimal places:

1)	$\ln 3.8$	[=1.335]		2)	$\ln 2.1$	[=0.7419]
3)	$\ln 7.2$	[=1.9741]		4)	$\ln 4.1$	[=1.4109]
5)	$\ln 1.5$	[=0.4055]		6)	$\ln 49$	[=3.8918]
7)	$\ln 36$	[=3.5835]		8)	$\ln 27$	[=3.2958]
9)	$\ln 183$	[=5.2095]		10)	$\ln 2.4612$	[=0.9006]
11)	$\exp 0.284$	[=1.3284]		12)	$\exp 3.14$	[=23.1039]
13)	$\exp 4$	[=54.5981]		14)	$\exp 3.142$	[=23.1501]
15)	$\exp 4.2184$	[=67.9247]				

Evaluate the following to five figures:

16)	$\exp 41$	[=6.3984×10^{17}]		17)	$\exp 32$	[=7.8963×10^{13}]
18)	$\exp 1.43$	[=4.1787]				

Evaluate to seven figures:

19)	$\exp 4.3$	[=73.76998]	20)	Evaluate to eight figures $\exp 11.08$
				[=64860.883]

Chapter 8

CHANGE OF ROOTS OF EQUATIONS

In this chapter we consider how to change a polynomial to obtain another whose roots are related to the original equation by being:

A) opposite in sign,
B) reciprocals,
C) multiples or sub-multiples,
or D) increased or reduced by a given number.

(a) ROOTS OPPOSITE IN SIGN

To obtain an equation whose roots are opposite in sign to those of a given polynomial equation.

We transpose the sign of every alternate term, beginning with the second.

Example 1 $x^4 - 2x^3 - 13x^2 + 38x - 24 = 0$ has roots 1, 2, 3, –4.
   $\therefore\ x^4 + 2x^3 - 13x^2 - 38x - 24 = 0$ has roots –1, –2, –3, 4.

Example 2 $2x^2 + 5x - 3 = 0$ has roots ½, –3
   $\therefore\ 2x^2 - 5x - 3 = 0$ has roots –½, 3

Proof   If $f(x) = 0$ is a polynomial equation and, and if $-y = x$, then $f(-y) = 0$ is satisfied by every root of $f(x) = 0$ with its sign transposed.
   $\therefore\ f(-y) = 0$ has roots opposite in sign to those of $f(x) = 0$, and since successive terms in any polynomial alternate between odd and even powers of the unknown, by substituting $-y$ for x effectively changes the sign of alternate terms.

(b) RECIPROCAL ROOTS

To obtain an equation whose roots are reciprocals of those of a given equation we reverse the order of the coefficients.

Fxample 3 $2x^3 + 3x^2 - 8x + 3 = 0$ has roots ½, 1, –3.

The coefficients are 2, 3, -8, 3.

$\therefore 3x^3 - 8x^2 + 3x + 2 = 0$ has roots 2, 1, $-\frac{1}{3}$.

Example 4 $15x^2 - 13x + 2 = 0$ has roots $\frac{2}{3}, \frac{1}{5}$.

$\therefore 2x^2 - 13x + 15 = 0$ has roots $\frac{3}{2}$, 5.

Proof   If $f(x) = 0$ is a polynomial equation, and if $x = \frac{1}{y}$, then $f\left(\frac{1}{y}\right) = 0$ will have roots reciprocal to those of $f(x) = 0$. But making the substitution $x = \frac{1}{y}$ will, on multiplying through by the appropriate value of y give a polynomial equation with reversed coefficients.

E.g. If $ax^2 + bx + c = 0$ then $\frac{a}{y^2} + \frac{b}{y} + c = 0$ becomes $a + by + cy^2 = 0$

i.e. $cy^2 + by + a = 0$.

The method in these first two sections both come under the Vedic sutra **TRANSPOSE AND APPLY**.

(c) MULTIPLE OR SUB-MULTIPLE ROOTS

To obtain an equation whose roots are all equal to those of a given equation multiplied by a given number, we multiply the coefficients of the given equation by successive powers of the given number, starting with zero power.

This comes under the **PROPORTIONATELY** sutra.

Example 5 $x^2 + 2x - 15 = 0$ has roots 3, –5.

Suppose, we want an equation with double those roots, i.e. 6 and –10. Multiply the coefficients by $2^0, 2^1, 2^2$:

$x^2 + 2 \times 2x - 2^2 \times 15 = 0$, or $x^2 + 4x - 60 = 0$ is the required equation.

Example 6 $x^2 + 2x - 15 = 0$ has roots 3, -5.

For an equation with roots 9, -15, i.e. 3 times the original roots:
$x^2 + 6x - 135 = 0$.

For an equation with roots 1½, -2½, i.e. half the original roots:
$x^2 + x - 3\frac{3}{4} = 0$, or $4x^2 + 4x - 15 = 0$

Example 7 $2x^3 - 3x^2 - 2x + 3 = 0$ has roots 1½, -1, 1.

An equation with twice these roots is:
$2x^3 - 6x^2 - 8x + 24 = 0$ i.e. multiply by 1, 2, 4, 8.
$\therefore x^3 - 3x^2 - 4x + 12 = 0$.

Example 8 $2x^3 - 3x^2 - 2x + 3 = 0$ as above

We can in fact start with any term and multiply by powers of 2 to the right and divide by powers of 2 to the left. Thus we can multiply these coefficients by ½, 1, 2, 4 to double its roots:

$\therefore x^3 - 3x^2 - 4x + 12 = 0$.

Example 9   Given $6x^2 - x - 15 = 0$ find an equation with 3 times the roots.

Multiply by $\frac{1}{3}$, 1, 3.
$\therefore 2x^2 - x - 45 = 0$.

Proof   If $f(x) = 0$ is a polynomial equation, and if $y = xn$, then $f\left(\frac{y}{n}\right) = 0$ will have roots n times those of $f(x) = 0$.
But if $f\left(\frac{y}{n}\right) = 0$ then $y^p \times f\left(\frac{y}{n}\right) = 0$ where p is the order of the polynomial.

E.g. if $ax^2 + bx + c = 0$ then $\frac{ay^2}{n^2} + \frac{by}{n} + c = 0$ or $ay^2 + nby + n^2c = 0$.

(d) ROOTS INCREASED OR REDUCED BY SOME VALUE

To increase or decrease the roots of a polynomial equation by a given number h, we substitute the transposed value of the given number (i.e. -h) into successive derivatives of the polynomial.

Example 10   Given $x^2 + 2x - 15 = 0$ (which has roots 3, -5) find a polynomial equation of the same order with roots 3 more than these, i.e. 6 and -2.

We have $h = 3$
and if $Q = x^2 + 2x - 15$ then $Q' = 2x + 2$
$\therefore$ $Q(-3) = 9 - 6 - 15 = -12$ and $Q'(-3) = -6 + 2 = -4$

So the required equation is $\underline{x^2 - 4x - 12 = 0}$

Proof   Let $Q = x^2 - (a+b)x + ab$, with roots a, b
$\therefore$ $Q(x-h) = (x-h)^2 - (a+b)(x-h) + ab$
$\qquad\qquad = x^2 + [-2h - (a+b)]x + h^2 + (a+b)h + ab$
$\qquad\qquad = x^2 + Q'(-h)x + Q(-h)$

And if $Q = acx^2 - (bc+ad)x + bd$, with roots $\frac{b}{a}, \frac{d}{c}$
then $Q(x-h) = acx^2 + Q'(-h)x + Q(-h)$

Example 11   Reduce the roots of $6x^2 - x - 15 = 0$ by 4

$\therefore$ $h = -4$ and $Q(4) = 6(4^2) - 4 - 15 = 77$
$\qquad\qquad$ and $Q'(4) = 12\times4 - 1 = 47$

$\therefore$ the required equation is $\underline{6x^2 + 47x + 77 = 0}$

Example 12   Decrease the roots of $x^3 + 6x^2 + 11x + 6 = 0$ by 2

$\therefore$ $h = -2$, $Q(2) = 8 + 24 + 22 + 6 = 60$
As $Q'(x) = 3x^2 + 12x + 11$

$\therefore$ $Q'(2) = 3\times2^2 + 12\times2 + 11 = 47$

And as $Q''(x) = 6x + 12$ $\therefore$ $Q''(2) = 6\times2 + 12 = 24$

So the required equation is $\underline{x^3 + 12x^2 + 47x + 60 = 0}$

Note that the coefficient obtained from the second derivative is divided by 2, see below.

Proof   Let $C = x^3 - (a+b+c)x^2 + (ab+bc+ac)x - abc$
This has roots a, b, c.
$\therefore$ $C(x-h) = x^3 - (a+b+c+3h)x^2 + [ab+bc+ac+2h(a+b+c)+3h^2]x$
$\qquad\qquad\qquad\qquad - [h^3+(a+b+c)h^2 + (ab+bc+ac)h + abc]$
$\qquad\qquad = x^3 + \frac{C''(-h)x^2}{2!} + C'(-h)x + C(-h)$

Example 13   Increase the roots of $2x^3 - 3x^2 - 2x + 3 = 0$ by 1

We have $h = 1$ and $Q(-1) = \mathbf{0}$

$Q'(x) = 6x^2 - 6x - 2, \quad \therefore \ Q'(-1) = \mathbf{10}$

And $Q''(x) = 12x - 6, \quad \therefore \ Q''(-1) = \mathbf{-18}$

Therefore the required equation is $\underline{2x^3 - 9x^2 + 10x = 0}$

Chapter 9

COSINE, SINE AND INVERSE TANGENT

By using the left to right methods of calculation described in Chapter 1 and the series expansions for sinx, cosx and inverse tanx, these functions can be evaluated for any angle. This is the subject of the present chapter, and in the next chapter we show how these same expansions can be used in reverse to evaluate inverse sines, inverse cosines and tangents.

COSINE

The series expansion for cosx is:

$$\cos x = 1 - \frac{x^2}{2!} + \frac{x^4}{4!} - \frac{x^6}{6!} + \frac{x^8}{8!} - \dots \quad \text{where x is in radians}$$

If we write this as $\cos x = 1 + A + B + C + D + \dots$
we see that $B = \frac{A^2}{6}$, $C = \frac{AB}{15}$, $D = \frac{AC}{28}$ etc.

This means that once we have obtained the value of $\frac{x^2}{2}$ the next term is obtained by squaring this and dividing by 6, the next by multiplying the two previous terms and dividing by 15, and so on.

Example 1   Find cosine 0.2 radians

We set up a chart as shown:

$1 - \frac{x^2}{2}$		$1 . 0\ \bar{2}$	row 1
$\frac{x^4}{24}$	6	$.0\ 0\ 0\ 0\ 6\ 6\ 6\ 6\ 6\ 6$	row 2
$-\frac{x^6}{720}$	15	$\bar{1}\ _3 1\ _3 1\ _3 1\ _3$	row 3
$\cos 0.2 =$		$1 . 0\ \bar{2}\ 0\ 0\ 6\ 6\ 5\ 7\ 7\ 7$	row 4

We require the figures in row 4 and these are the sum of the figures in rows 1, 2 and 3, as can be seen by looking at the left-hand side. As explained above the divisors in rows 2 and 3 are 6 and 15 and it may be useful to have these written down.

We put 1 in row 1, followed by $\frac{-0.2^2}{2} = 0.0\bar{2}$

Then square the decimal part of row 1 and divide by 6 to obtain row 2:

$0.02^2 = 0.0004; \quad 0.0004 \div 6 = 0.00006$

Then multiply rows 1 and 2 and divide by 15:
$\bar{2} \times 6 = \bar{12}; \; \bar{12} \div 15 = \bar{1} \, r3;$ put down $\bar{1}_3$

then $\bar{2} \times 6 = \bar{12}; \; \bar{12} +$ carried 3 (as 30) = 18; 18 ÷ 15 = 1 r3; etc.

We then add the columns up. In fact the next term in the series comes in at the tenth decimal place, so that the last figure should be 8.

∴ cos0.2 = 0.980066578 to 9 D.P.

Example 2   Cosine 0.3 radians

This is very similar:

$1 - \frac{x^2}{2}$		$1 . 0 \; \bar{4} \; \bar{5}$							row 1
$\frac{x^4}{24}$	6		$3 \; _{\bar{2}}3 \; {}_27 \; {}_3 \, 5$						row 2
$-\frac{x^6}{720}$	15			$\bar{1} \; {}_30 \; {}_3\bar{1} \; {}_2$					row 3
cos0.3 =		$1 . 0 \;\; \bar{4} \;\; \bar{5} \;\; 3 \;\; 3 \;\; 6 \;\; 5 \;\; \bar{1}$							row 4
cos0.3 =		$0 . 9 \;\; 5 \;\; 5 \;\; 3 \;\; 3 \;\; 6 \;\; 4 \;\; 9$						to 8 D.P.	

Row 1: $\frac{-0.3^2}{2} = 0.0\bar{4}\bar{5}$

Row 2:   we need to square $\bar{4}5$, so we find the duplexes in turn, and divide each of them by 6: $D(\bar{4}) = 16; \; 16 \div 6 = 3 \, r\bar{2},$ put $3_{\bar{2}}.$

   $D(\bar{4}5) = 40; \; 40 +$ carried $\bar{2}$ (as $\bar{20}$) = 20; 20 ÷ 6 = 3 r2, put $3_2.$

   $D(\bar{5}) = 25, \; 25 + 20 = 45, \; 45 \div 6 = 7_3.$
   The carried 3 is 30 in the next column, so 30 ÷ 6 = 5.

Row 3:   we multiply rows 1, 2: $CP\begin{pmatrix} \bar{4} \\ 3 \end{pmatrix} = \bar{12}, \; \bar{12} \div 15 = \bar{1}_3$

   $CP\begin{pmatrix} \cdot\bar{4} & \bar{5} \\ 3 & 3 \end{pmatrix} = \bar{27}, \; \bar{27} + 30 = 3, \; 3 \div 15 = 0_3$

$$CP\begin{pmatrix} \overline{4} & \overline{5} \\ 3 & 7 \end{pmatrix} = \overline{43}, \ \overline{43} + 30 = \overline{13}, \ \overline{13} \div 15 = \overline{1}_2$$

Finally adding the columns and removing the bar figures gives the answer.

Example 3   Cos0.1982

$$\cos 0.1982 = \cos 0.20\overline{2}2$$

$1 - \frac{x^2}{2}$		1 . 0 $\overline{2}$ 0 4 $\overline{4}$ $\overline{2}$ 4	row 1
$\frac{x^4}{24}$	6	1 $\overline{2}$ $\overline{3}$ $\overline{2}$ $\overline{6}$ 0 3 $\overline{2}$	row 2
$-\frac{x^6}{720}$	15	$\overline{1}$ $_1$	row 3
$\cos 0.20\overline{2}2 =$		1 . 0 $\overline{2}$ 0 5 $\overline{7}$ $\overline{8}$ 6	row 4

$$\therefore \ \cos 0.1982 = \ 0 \ . 9 \ \ 8 \ \ 0 \ \ 4 \ \ \ 2 \ \ \ 2 \ \ \ 6$$

To obtain the figures of row 1 we take successive duplexes of $0.20\overline{2}2$ divide by 2 and change the sign:

D(2) = 4, 4 ÷ 2 = 2, put down $\overline{2}$;   D(20) = 0, 0 ÷ 2 = 0, put down 0;

D(20$\overline{2}$) = 8, 8 + 2 = 4, put down 4;   D(2022) = 8, 8+2 = 4, put down $\overline{4}$;

D(0$\overline{2}$2) = 4, 4+2 = 2, put down $\overline{2}$;   D(22) = 8, 8 + 2 = 4, put down 4.

For row 2, take duplexes in row 1 and divide by 6:

D($\overline{2}$) = 4, 4÷6 = 1$\overline{2}$;   D($\overline{2}$0) = 0, 0+$\overline{2}$0 = $\overline{2}$0, $\overline{2}$0 + 6 = 3$\overline{2}$;

D($\overline{2}$04) = $\overline{16}$, $\overline{16}$ + $\overline{2}$0 = $\overline{36}$, $\overline{36}$ + 6 = $\overline{6}$0 , D($\overline{2}$044) = 16, 16+0=16, 16÷6 = 3$\overline{2}$;

For row 3, take cross-products of rows 1 and 2 and divide by 15:

$$CP\begin{pmatrix} \overline{2} \\ 1 \end{pmatrix} = \overline{2} \text{ and } CP\begin{pmatrix} \overline{2} & 0 \\ 1 & 3 \end{pmatrix} = 6, \text{ combining these gives } \overline{26} = \overline{14}, \ \overline{14} \div 15 = \overline{1}_1$$

Next add rows 1, 2, 3 to get cos0.1982 = 0.9804226 to 7 D.P.

SINE

The series expansion for sinx is:

$$\sin x = x - \frac{x^3}{3!} + \frac{x^5}{5!} - \frac{x^7}{7!} + \ldots$$

Writing this as $\sin x = A - B + C - D + \ldots$

we see that $B = \frac{Ax^2}{6}$, $C = \frac{Bx^2}{20}$, $D = \frac{Cx^2}{42}$ etc.

This means that we can obtain x^2 and then evaluate each term by multiplying by x^2 and dividing by 6, 20, 42 etc.

Example 4   Sin0.2

$(x^2$	0 . 0 4)		row 0
x	0 . 2		row 1
$-\frac{x^3}{3!}$ 6	. 0 0 $\bar{1}_2$ $\bar{3}_2$ $\bar{3}_2$ $\bar{3}_2$ $\bar{3}_2$ $\bar{3}_2$ $\bar{3}_2$		row 2
$\frac{x^5}{5!}$ 2	2 6 6 6		row 3
$-\frac{x^7}{7!}$ 42	$\bar{2}_{20}$		row 4
$\sin 0.2 =$	0 . 2 0 $\bar{1}$ $\bar{3}$ $\bar{3}$ $\bar{1}$ 3 3 1		row 5

First we find that $x^2 = 0.04$

For row 2, we multiply rows 0 and 1 and divide by 6:
$4 \times 2 = 8$, $8 \div 6 = 1$ r2, put down 1_2 (we will change the signs of rows 2 and 4 later).
Take the carried 2; as 20, $20 \div 6 = 3_2$ etc.

For row 3, multiply rows 0 and 2 and divide by 2:
$4 \times 1 = 4$, $4 \div 2 = 2$
$4 \times 3 = 12$, $12 \div 2 = 6$, etc.

For row 4, multiply rows 0 and 3 and divide by 42:
$4 \times 26 = 104$, $104 \div 42 = 2_{20}$

Then change the signs of rows 2 and 4 and add the columns up.

$\therefore$ $\sin 0.2 = 0.19866933$ to 8 D.P.

Example 5   Sin0.3

$(x^2$	0 . 1 $\bar{1}$)
x	0 . 3
$-\frac{x^3}{3!}$ 6	. 0 0 $\bar{4}_3$ $\bar{5}$
$\frac{x^5}{5!}$ 2	2 0$_1$ 2 5
$-\frac{x^7}{7!}$ 42	$\bar{4}$
$\sin 0.3 =$	0 . 3 0 $\bar{4}$ $\bar{5}$ 2 0 2 1

Example 6   Sin 0.23 .

Here we have 2 digits after the decimal point.

$$
\begin{array}{llll}
(x^2 & \quad .0 \;\; 5 \;\; 3 \;\; \overline{1}) & \text{row 0} \\
x & \quad 0 \; . \; 2 \;\; 3 & \text{row 1} \\
-\frac{x^3}{3!} \quad 6 & \quad .\; 0 \;\; 0 \;\; \overline{2}_{\overline{2}} \; 0_1 \; \overline{3}_{\overline{1}} \; \overset{+}{2}_{\overline{1}} \overset{+}{2}_{2} \overline{3} \;\; \overline{3} & \text{row 2} \\
\frac{x^5}{5!} \quad 20 & \qquad\qquad\quad\; 5 \;\; 3 \;\; 6_1 \; 4_1 & \text{row 3} \\
-\frac{x^7}{7!} \quad 42 & \qquad\qquad\qquad\qquad\quad\; 7_{\overline{14}} & \text{row 4} \\
\hline
\sin 0.23 = & \;\; 0 \; . \; 2 \;\; 3 \;\; \overline{2} \;\; 0 \;\; \overline{3} \;\; 7 \;\; 5 \;\; 3 \;\; \overline{6}
\end{array}
$$

For row 2 we multiply rows 0 and 1 and divide by 6:

$$CP\binom{5}{2} = 10, \;\; 10 + 6 = 2_{\overline{2}}$$

$$CP\binom{5 \;\; 3}{2 \;\; 3} = 21, \;\; 21 + \overline{20} = 1, \;\; 1 + 6 = 0_1$$

$$CP\binom{3 \;\; \overline{1}}{2 \;\; 3} = 7, \;\; 7 + 10 = 17, \;\; 17 + 6 = 3_{\overline{1}}$$

$$CP\binom{\overline{1}}{3} = \overline{3}, \;\; \overline{3} + \overline{10} = \overline{13}, \;\; \overline{13} + 6 = 2_{\overline{1}}$$

For row 3 multiply rows 0 and 2 and divide by 2:

$$CP\binom{5}{2} = 10, \;\; 10 + 2 = 5$$

$$CP\binom{5 \;\; 3}{2 \;\; 0} = 6, \;\; 6 + 2 = 3$$

$$CP\binom{5 \;\; 3 \;\; \overline{1}}{2 \;\; 0 \;\; 3} = 13, \;\; 13 + 2 = 6_1$$

$$CP\binom{5 \;\; 3 \;\; \overline{1}}{0 \;\; 3 \;\; \overline{2}} = \overline{1}, \overline{1} + 10 = 9, \;\; 9 + 2 = 4_1$$

For row 4: $CP\binom{5}{5} = 25$ and $CP\binom{5 \;\; 3}{5 \;\; 3} = 30$, together these give 280,
$280 + 42 = 7_{\overline{14}}$

$\therefore \sin 0.23 = 0.22797752$ to 8 D.P.

The solution converges well for small angles. For larger angles we can always use the nearest quadrant boundary. This means that the most awkward cases are when the angle is near $\frac{\pi}{4}$ radians. So we tackle one more sine where the angle is close to this- and also has four figures after the decimal point.

Example 7   Sin0.7132

$$
\begin{array}{l}
(x^2 \qquad\quad 0\ .\ 5\ _{\bar{1}}\ 1\ _{\bar{6}}\ \bar{1}\ _{\bar{7}}\ \bar{3}\ _{\bar{6}}\ \bar{5}\ _3\ 4\ _2) \\
x \qquad\qquad 0\ .\ 7\quad 1\quad 3\quad 2 \\
-\frac{x^3}{3!}\quad 6 \qquad\quad \bar{6}\ _{\bar{1}}\ 0\ _2\ \bar{5}\ _{\bar{1}}\ 3\ _{\bar{1}}\ \overset{+}{8}\ _{\bar{1}} \\
\frac{x^5}{5!}\quad 20 \qquad\qquad 1\ _{10}\ 5\ _6\ 4\ _{\bar{1}}\ \overset{+}{2}\ _2 \\
-\frac{x^7}{7!}\quad 42 \qquad\qquad\qquad \bar{2}\ _{\bar{8}}\ \overset{+}{1}\ _{\overline{14}}
\end{array}
$$

$$\sin 0.7132 = \ 0\ .\ 7\ \ \bar{5}\ \ 4\ \ 2\ \ 5\ \ 7$$

We have to be careful if we want to avoid large figures here.
The squaring in the first line is similar to that in Example 69, page 14.

INVERSE TANGENT

The series is: $\tan^{-1}x = x - \frac{x^3}{3} + \frac{x^5}{5} - \frac{x^7}{7} + \ldots$

Example 8   Find $\tan^{-1}0.2$

$$
\begin{array}{l}
x \qquad\qquad 0\ .\ 2 \\
-\frac{x^3}{3} \qquad\quad .\ 0\quad 0\ \ \bar{3}\quad 3\quad 3\quad 3\quad 3 \\
\frac{x^5}{5} \qquad\qquad\qquad\qquad\qquad 6\quad 4 \\
-\frac{x^7}{7} \qquad\qquad\qquad\qquad\qquad\qquad \bar{2}\quad 2
\end{array}
$$

$$\tan^{-1}0.2 = \ 0\ .\ 2\ \ 0\ \ \bar{3}\ \ 3\ \ 9\ \ 5\ \ 5$$

Here it is very easy to evaluate the terms and add them up.

Example 9   Tan$^{-1}0.342$

x	0 . 3	4	2			row 1
$-\frac{x^3}{3}$		$\bar{1}$	$\bar{3}$ $_1$	$\bar{3}$ $_2$		row 2
$\frac{x^5}{5}$			1 $_{\bar{2}}$	$\bar{1}$ $_1$		row 3
$\tan^{-1}0.342 =$	0 . 3	3	0	$\bar{4}$		row 4
$(x^2$	1 $_{\bar{1}}$	2 $_{\bar{6}}$	$\bar{3}$ $_{\bar{2}}$ $)$			row 5
$(x^4$	1	4	$)$			row 6

We first find x^2 in row 5.

For row 2 we multiply rows 1 and 5 and divide by 3:

$$CP\begin{pmatrix}3\\1\end{pmatrix} = 3,\ 3+3 = 1$$

$$CP\begin{pmatrix}3&4\\1&2\end{pmatrix} = 10,\ 10+3 = 3_1$$

$$CP\begin{pmatrix}3&4&2\\1&2&3\end{pmatrix} = 1,\ 1+10 = 11,\ 11+3 = 3_2$$

For row 3, square row 5 (row 6), multiply by row 1, and divide by 5:

$$D(1) = 1,\ CP\begin{pmatrix}3\\1\end{pmatrix} = 3,\ 3+5 = 1\overline{5}$$

$$D(12) = 4,\ CP\begin{pmatrix}3&4\\1&4\end{pmatrix} = 16,\ 16+\overline{20} = \overline{4},\ \overline{4}+5 = \overline{1}_1$$

Change the signs of the figures in row 2, and add up the columns.

□ $\tan^{-1}0.342 = 0.3296$

Example 10 $\tan^{-1}20$

This angle is close to 90°, and, as can be seen from the diagram,

we can find it by first finding $\tan^{-1}\frac{1}{20}$ and then subtracting the result from $\frac{\pi}{2}$.

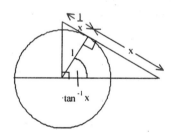

X	0 . 0 5					
$-\frac{x^3}{3}$		0	0	0 0 $\overline{4}$	$\overline{2}$	
$\tan^{-1}0.05 =$	0 . 0	4	9	9	5	8

□ $\tan^{-1}20 = 1.570796 - 0.049958 = 1.520838$

Example 11   Sin1.331

$$\sin 1.331 = \cos(\tfrac{\circ}{2} - 1.331) = \cos 0.2398 = \cos 0.2402 :$$

$1-\frac{x^2}{2}$	20	1 . 0 $_4$ $\overline{3}$ $_{\overline{4}}$ $\overline{1}$ $_{\overline{4}}$ 2 $_8$ 5 4 $\overline{2}$ 0 0 $\overline{2}$
$\frac{x^4}{24}$	6	1 3 4 0 $\overline{2}$ $_1$ 3 2 8 $_{\overline{2}}$
$-\frac{x^6}{720}$	15	$\overline{3}$ 4 4 $_{\overline{8}}$
$\sin 1.331 =$		1 . 0 $\overline{3}$ 1 3 9 $\overline{4}$ $\overline{5}$ 0

☐ $\sin 1.331 = 0.97138550$

In the first line of this calculation we have used 20 as a divisor because the duplexes obtained give some large numbers if divided by 2.

For the first line:
$D(2) = 4$, $4 \div 20 = 0$ r4, put 0_4
$D(24) = 16$, $16 + 40 = 56$, $56 \div 20 = 3_{\overline{4}}$
$D(240) = 16$, $16 - 40 = \overline{24}$, $\overline{24} \div 20 = \overline{1}_{\overline{4}}$
$D\left(240\overline{2}\right) = \overline{8}$, $\overline{8} - 40 = \overline{48}$, $\overline{48} \div 20 = \overline{2}_{\overline{8}}$
$D\left(40\overline{2}\right) = \overline{16}$, $\overline{16} - 80 = \overline{96}$, $\overline{96} \div 20 = \overline{5}_4$
etc.

HYPERBOLIC FUNCTIONS

These can also be expressed simply in series form, and so can be evaluated like the previous examples.

Example 12   Sinh0.41

We have $\sinh x = x + \frac{x^3}{3!} + \frac{x^5}{5!} + \frac{x^7}{7!} + \dots$

x		0 . 4	1				
$\frac{x^3}{3!}$	6			$1_2\ 2_{\overline{2}}\ \overline{5}_{\overline{1}}$	$\overline{1}_{\overline{2}}\ \overline{3}_{\overline{1}}$		
$\frac{x^5}{5!}$	2				$1_0\ 0_1\ \overline{4}_0$		
$\sinh 0.41 =$		0 . 4	2	2	$\overline{4}$	$\overline{1}$	7

☐ $\sinh 0.41 = 0.42158$ to 5 D.P.

Similarly we can evaluate $\cosh x$ and $\tanh^{-1} x$ using:

$$\cosh x = 1 + \frac{x^2}{2!} + \frac{x^4}{4!} + \frac{x^6}{6!} + \dots$$

$$\tanh^{-1} x = x + \frac{x^3}{3} + \frac{x^5}{5} + \frac{x^7}{7} + \dots$$

However hyperbolic functions are not periodic like the circular functions, so that if x is large a lot of work would be involved. But in such cases we can use the methods described in Chapter 7 to evaluate the following alternative expressions for sinhx, coshx and tanhx:

$$\sinh x = \tfrac{1}{2}(e^{-x} - e^x) \qquad \cosh x = \tfrac{1}{2}(e^{-x} + e^x) \qquad \tanh x = \frac{e^{2x}-1}{e^{2x}+1}$$

Chapter 10

INVERSE SINE, COSINE AND TANGENT

Having the ability to make calculations from left to right means that we can obtain the most significant digit of an answer first, then the next most significant, and so on. In this chapter and the next we show how the digits of an answer, as they appear, can be used to obtain the next digit.

INVERSE SINE

Since $\sin \underline{x} = x - \frac{x^3}{3!} + \frac{x^5}{5!} - \frac{x^7}{7!} + \ldots$

$\square \ x = \sin x + \frac{x^3}{3!} - \frac{x^5}{5!} + \ldots .$

Example 1   Find $\sin^{-1} 0.4$ to 4 D.P.

Here we are given $\sin x = 0.4$, and we have to find x.

So $x = 0.4 + \frac{x^3}{3!} - \frac{x^5}{5!} + \ldots$

We therefore set up a chart to evaluate the terms on the RHS of this equation:

$\sin x$		0 . 4							row 1
$\frac{x^3}{3!}$	6		1_2	1_4	$6_{\bar{1}}$	1_2			2
$-\frac{x^5}{5!}$	2				$\bar{1}$	$0_{\bar{1}}$			3
$x =$		0 . 4	1	1	5	1			4
$(x^2=$			$2_{\bar{4}}$	3_2	$1_{\bar{1}}$	$3_2)$			5

The answer appears in row 4, and rows 2, 3 and 5 are formed digit by digit as x is found, digit by digit. We therefore work vertically (rather than horizontally as in the last chapter) and as each digit of x is found it is used to find the next one.

First bring 0.4 down into the answer. Then for row 5, D(4) = 16 = $2\bar{4}$, put $2_{\bar{4}}$

Next, for row 2, multiply rows 4 and 5 and divide by 6:

$$CP\left(\begin{array}{c} 4 \\ 2 \end{array}\right) = 8, \ 8+6 = 1_2.$$

Bring 1 down into the answer; then for row 5, D(41) = 8, $8+\overline{40} = \overline{32}$. put $3_{\bar{2}}$

Row 2 again: $CP\left(\begin{array}{cc} 4 & 1 \\ 2 & 3 \end{array}\right) = \overline{10}, \ \overline{10}+20 = 10, \ 10+6 = 1_4$

Bring down 1 into the answer.

Row 5: D(411) = 9, $9 + \overline{20} = \bar{1}_{\bar{1}}$

Row 2: $CP\left(\begin{array}{ccc} 4 & 1 & 1 \\ 2 & 3 & 1 \end{array}\right) = \bar{5}, \ \bar{5} + 40 = 35, \ 35+6 = 6_{\bar{1}}$

Row 3: we multiply rows 2 and 5 and divide by 2:

$$CP\left(\begin{array}{c} 1 \\ 2 \end{array}\right) = 2, \ 2+2 = 1, \ \text{put } \bar{1} \text{ as this term is negative.}$$

Bring down 5 into the answer.

Row 5: D(4115) = 42, $42 + \overline{10} = 3_2$

Row 2: $CP\left(\begin{array}{cccc} 4 & 1 & 1 & 5 \\ 2 & 3 & 1 & 3 \end{array}\right) = 18, \ 18 + \overline{10} = 8, \ 8+6 = 1_2$

Row 3: $CP\left(\begin{array}{cc} 1 & 1 \\ 2 & 3 \end{array}\right) = \bar{1}, \ \bar{1}+2 = 0_{\bar{1}}$

Bring down 1 into the answer.

$\therefore \ x = 0.4115$ to 4 D.P.   (the other solution between 0 and 2π is $\pi - 0.4115$)

Example 2   Find $\sin^{-1} 0.2195$

Similarly we want to solve: $\sin x = 0.220\bar{5}$:

$\sin x$		0 . 2	2	0	$\bar{5}$			
$\frac{x^3}{3!}$	6		1_4	8	0_3	6_2		
$-\frac{x^5}{5!}$	2				$\bar{4}_9$			
$x \ =$		0 . 2	2	1	3	0	2	
($x^2=$		$.0_{\ 45\bar{2}}$	$1_{\bar{2}}$	$0_{\bar{4}}$	3_3)			

The procedure is very similar to that of the first example.

It is prudent in all these examples to have an eye on the next column, duplex or cross-product when deciding what quotient and remainder to put down.

It should also be noted that the particular figures chosen in the working of these examples are not the only ones that could be used.

INVERSE COSINE

Example 3   Find $\cos^{-1} 0.123$

Since this angle is close to 90°, we can find $\sin^{-1} 0.123$ and subtract the result from $\frac{\pi}{2}$:

$\sin x$	$0 . 1 \quad 2 \quad 3$	
$\frac{x^3}{3!}$ $\quad 6$	$3_{\bar{1}} \; 1_{\bar{1}} \; 2_0$	
$x =$	$0 . 1 \quad 2 \quad 3 \quad 3 \quad 1 \quad 2$	
$(x^2 =$	$.0\,0_{\,1}1_4 \; 5_0 \; 2_{\bar{2}} \; 0_{\,3})$	

$+ \;\; \cos^{-1} 0.123 = \frac{\pi}{2} - 0.123312 = 1.447484$

Example 4   Find $\cos^{-1} 0.8$

Since $\cos x = 1 - \frac{x^2}{2!} + \frac{x^4}{4!} - \frac{x^6}{6!} + \frac{x^8}{8!} - \ldots$

$\square \;\; x^2 = 2(1 - \cos x) + \frac{x^4}{12} - \frac{x^6}{360} + \ldots .$

And if we let $y = x^2$ we have: $y = 2(1 - \cos x) + \frac{y^2}{12} - \frac{y^3}{360} + \ldots .$

so that we can evaluate y and take its square root (see page 16 et seq) to find x.

So if $\cos x = 0.8$, then $y = 2(1 - 0.8) + \frac{y^2}{12} - \frac{y^3}{360} + \ldots .$

$2(1 - \cos x)$	$0 . 4$						row 1
$\frac{y^2}{12}$ $\quad 12$		$1_4 \; 4_0 \; 3_{\bar{3}} \; \bar{1}_{\bar{2}}$					2
$-\frac{y^3}{360}$ $\quad 3$		$2_{\bar{2}} \; 0_{\bar{3}}$					3
$y = x^2 =$	$0 . 4$	$1_5 \; 4_6 \; 1_9 \; \bar{1}_5$					4
$x =$	$0 . 6$	$4 \quad 3 \quad 5$					5

First we bring 0.4 down into row 4.
Row 2: D(4) = 16, 16 ÷ 12 = 1_4

Bring down 1 into row 4
Row 2: D(41) = 8, 8 + 40 = 48, 48 ÷ 12 = 4_0

Bring down 4 into row 4
Row 2: D(414) = 33, 33 + 0 = 33, 33 ÷ 12 = $3_{\bar{3}}$

Row 3: multiply rows 2 and 4 and divide by 3:

$$CP\begin{pmatrix} 1 \\ 4 \end{pmatrix}4, \quad 4+3 = 2\ r\bar{2},\ \text{put } \bar{2}_{\bar{2}} \text{ as this term is negative}$$

Bring down 1 into row 4
Row 2: D(4141) = 16, 16 + $\overline{30}$ = $\overline{14}$, $\overline{14}$ ÷ 12 = $\bar{1}_{\bar{2}}$

Row 3: $CP\begin{pmatrix} 1 & 4 \\ 4 & \bar{1} \end{pmatrix}$ = 1.7, $17+\overline{20} = \bar{3}$, $\bar{3} + 3 = 0_{\bar{3}}$

Bring $\bar{1}$ into row 4
We then find the square root of row 4 using the usual Vedic method.

Alternatively we can find $\sin^{-1}0.6$ and subtract this from $\frac{\pi}{2}$.

INVERSE TANGENT

Example 5   Tan0.3

We have $\tan^{-1}x = x - \frac{x^3}{3} + \frac{x^5}{5} - \ldots$

□ $x = \tan^{-1}x + \frac{x^3}{3} - \frac{x^5}{5} + \ldots$

$\tan^{-1}x$	0 . 3				row 1
$\frac{x^3}{3}$ 3		1_0	0_1 $\bar{1}_0$		2
$-\frac{x^5}{5}$ 5			$\bar{1}_{\bar{2}}$ $\overset{+}{4}_0$		3
x =	0 . 3	1	$\bar{1}$	3	4
(x^2=	. $1_{\bar{1}}$	$0_{\bar{4}}$	$\bar{4}_{\bar{5}}$)		5

Put 0.3 in the answer.
Row 5: D(3) = $1_{\bar{1}}$

Row 2: $\text{CP}\begin{pmatrix} 3 \\ 1 \end{pmatrix} = 3, 3 + 3 = 1_0$, put 1 in the answer.

Row 5: $\text{D}(31) = 6, 6 + \overline{10} = \overline{4}$, put $0_{\overline{4}}$

Row 2: $\text{CP}\begin{pmatrix} 3 & 1 \\ 1 & 0 \end{pmatrix} = 1, 1+0=1, 1+3 = 0_1$

Row 3: multiply rows 2 and 5, multiply by 3 and divide by 5:

$$\text{CP}\begin{pmatrix} 1 \\ 1 \end{pmatrix} = 1, 1.3 = 3, 3 + 5 = 1_{\overline{2}}$$

Put $\overline{1}$ in the answer.
Row 5: $\text{D}(31\overline{1}) = \overline{5}, \overline{5} + \overline{40} = \overline{4}_{\overline{5}}$

Row 2: $\text{CP}\begin{pmatrix} 3 & 1 & \overline{1} \\ 1 & 0 & 4 \end{pmatrix} = \overline{13}, \overline{13} + 10 = \overline{3}, \overline{3} + 3 = \overline{1}_0$

Row 3: $\text{CP}\begin{pmatrix} 1 & 0 \\ 1 & 0 \end{pmatrix} = 0, 0.3=0, 0+\overline{20} = \overline{20}, \overline{20} + 5 = \overline{4}_0$ *

Put 3 in the answer.

*Note that we multiply by 3 and add the carried figure **before** dividing by 5.

Example 6   Tan1.248

1.248 radians is too large for us to handle easily, so we go for the complementary angle, as we did in Example 10 in the last chapter, find its tangent and take the reciprocal.

$\tan(\frac{\circ}{2} - 1.248) = \tan 0.3228$

$\tan^{-1}x$		0 .	3	2	3	$\overline{2}$	0
$\frac{x^3}{3}$	3			1_0	2_0	$5_{\overline{2}}$	$\overline{2}_{\overline{2}}$
$-\frac{x^5}{5}$	5				$\overline{1}_{\overline{2}}$	$\overset{+}{2}_{\overline{1}}$	$\overline{4}_{\overline{3}}$
$\frac{x^7}{7}$	7						6_3
x =		0 .	3	3	4	5	0
(x²=		.	$1_{\overline{1}}$	$1_{\overline{2}}$	$2_{\overline{7}}$	$\overline{1}_{\overline{6}}$)	

Then $1 \div 0.33450 = 2.990$, using the reciprocal method from Chapter 1 or straight division.

☐ tan 1.248 = 2.990

Example 7   Tan 1.884

Here we observe
that the tangent
has a negative value,
and, as can be seen
from the diagram,
we need to evaluate
$\tan\left(1.884-\tfrac{\pi}{2}\right)$, take
the reciprocal and
change the sign:

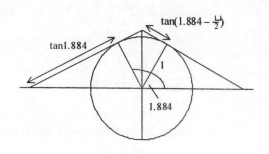

$$
\begin{array}{c|cccccc}
\tan^{-1}x & 0 & . & 3 & 1 & 3 & 2 & 0\\
\frac{x^3}{3} \quad 3 & & & & 1_0 & 1_{\bar1} & 3_{\bar1} & 2_{\bar1}\\
-\frac{x^5}{5} \quad 5 & & & & & \bar1_{\bar2} & 3_{\bar2} & 1_{\bar1}\\
\frac{x^7}{7} \quad 7 & & & & & & & 5_0\\
\hline
x \;=\; & 0 & . & 3 & 2 & 3 & 8 & 6\\
(x^2= & & . & 1_{\bar1} & 0_2 & 5_{\bar8} & \bar1_{\overline{10}})
\end{array}
$$

$= \tan(1.884 - \tfrac{\square}{2})$

$\square$ $\tan 1.884 = -1 + 0.32386 = -3.088$

HYPERBOLIC FUNCTIONS

We can similarly find inverse hyperbolic sines, inverse hyperbolic cosines and tangents using the series expansions given in the last chapter. The method is just like that of the previous examples of this chapter, so we give just one example, without explanations.

Example 8   Find $\sinh^{-1} 0.5432$

$\square$ $0.5432 = x + \frac{x^3}{3!} + \frac{x^5}{5!} + \dots$

$\square$ $x = 0.5432 - \frac{x^3}{3!} - \frac{x^5}{5!} - \dots$

$$
\begin{array}{c|cccccc}
\sinh x & 0 & . & 5 & 4 & 3 & 2\\
-\frac{x^3}{3!} \quad 6 & & & & \bar2_3 & \bar3_3 & \bar4_0 & \overset{+}{3}_{\bar2}\\
\frac{x^5}{5!} \quad 2 & & & & & & \bar3_0 & \bar1_{\bar1}\\
\hline
x \;=\; & 0 & . & 5 & 2 & 0 & \bar5 & 2\\
(x^2= & & . & 3_{\bar5} & \bar3_0 & 0_4 & \bar1_0)
\end{array}
$$

POLYNOMIAL EQUATIONS

In all these examples we have effectively been solving polynomial equations because we have discarded all the terms of the series after a certain point. In the example above we solved a quintic equation. We can generally solve polynomial equations in this way provided that the root we are looking for lies between 1 and −1. If it does not, then we can transform the equation, using the methods described in Chapter 8, so that the root is in this range, and then solve it. The following are a few examples of this, first without and then with change of roots.

Example 9 $x^2 + 5x - 1 = 0$

$$\therefore x = \frac{1 - x^2}{5}$$

We can see that this equation has a root at about $x = 0.2$.
We therefore solve $x = 0.2 - \frac{x^2}{5}$:

0.2	0 . 2				
$-\frac{x^2}{5}$		$\bar{1}_{\bar 1}$ 3_1 $\bar 4_3$ $\bar 2_{\bar 2}$			
x =	0 . 2	$\bar 1$ 3 $\bar 4$ $\bar 2$			

Bring down 2 into the answer.
Then $D(2) = 4$, $4 \div 5 = 1$ r$\bar 1$, put $1_{\bar 1}$

Bring down $\bar 1$, then $D(2\bar 1) - \bar 4$, $\bar 4 + \bar{10} = \bar{14}$, $\bar{14} \div 5 = \bar 3$ r1, put 3_1
And so on.

$$\therefore x = 0.19258 \text{ to 5 D.P.}$$

Example 10   Solve $2x^3 - 4x^2 - 10x + 3 = 0$

If we rearrange this equation to get $x = 0.3 - \frac{2x^2}{5} + \frac{x^3}{5}$

we see that there is a root at about $x = 0.3$.
We therefore proceed as before:

0.3	0 . 3						
$-\frac{2x^2}{5}$		$\bar 3_3$ $0_{\bar 6}$ 0_6 $\bar 5_{\bar 1}$					1
$\frac{x^3}{5}$		4_1 $1_{\bar 1}$ $2_{\bar 3}$					2
x =	0 . 3	$\bar 3$ 4 1 $\bar 3$					3
							4

Bring down 0.3 into the answer.
We need to square x, multiply by 2, divide by 5 and change the sign:

$D(3) = 9, 9 \times 2 = 18, 18 \div 5 = 3$ r3, put $\overline{3}_3$

Bring down $\overline{3}$

$D(3\overline{3}) = \overline{18}, \overline{18}\%2 = \overline{36}, \overline{36} + 30 = \overline{6}, \overline{6} + 5 = 0_{\overline{6}}$

Row 3: we change the sign of the figures in row 2, multiply by x (i.e. by row 4), and divide by 2:

$$CP\binom{3}{3} = 9,\ 9 + 2 = 4_1$$

Etc.

Alternatively we can evaluate $\quad x = 0.3 - 0.4x^2 + 0.2x^3$:

0.3	0 . 3						
$-0.4x^2$		$\overline{3}_6$ $0_{\overline{12}}$ $0_{12}\overline{5}_{\overline{2}}$ $\overset{+}{4}_{\overline{12}}$					
$0.2x^3$		4_1 $1_{\overline{1}}$ $2_{\overline{2}}$ $\overline{3}_4$					
x =	0 . 3	$\overline{3}$ 4 1 $\overline{3}$ 1					

Bring down 0.3 into the answer.

$D(3) = 9, 9 \times 4 = 36$, put $\overline{3}_6$

Bring down $\overline{3}$
$D(3\overline{3}) = \overline{18}, \overline{18} \times 4 = \overline{72}, \overline{72} + 60 = \overline{12}$, put $0_{\overline{12}}$ etc.
Row 3 is found as before.

$\therefore x = 0.274071$ to 6 D.P.

Example 11   Solve $7x^2 - 50x + 8 = 0$

This can be rearranged to $x = 0.16 + \frac{7x^2}{50}$ so that there is a root near $x = 0.2$.

0.16	0 . 2	$\overline{4}$			
$\frac{7x^2}{50}$		0_{28} 3 $_{18}$ 7 $_{26}$ 5 $_{38}$			
x =	0 . 2	$\overline{4}$ 3 7 5			

The other root is not less than 1, but by transforming the equation to another with reciprocal roots, we will be able to find it:

The transformed equation is $8y^2 - 50y + 7 = 0$ (see Chapter 8, section B)

$$\therefore y = \frac{7+8y^2}{50} = 0.14 + \frac{4y^2}{25}$$

0.14	0	.	1	4			
$\frac{4y^2}{25}$					$3\,\bar{3}$	$2\,8\,8\,\bar{8}$	
y =	0	.	1	4	3	2	8

We can then find the reciprocal of this using the Vedic method of straight division, see page 10 (or using the method shown on page 13):

$$3\ 3\ \bar{2})\ 1\ \ 0\ \ 0\ .\ 0\ \ 0\ \ 0\ \ 0$$
$$14 \qquad\qquad\quad \bar{4}\ \ \bar{1}\ \ \bar{4}\ \ \bar{1}\ \ 4$$
$$\overline{\qquad\qquad 1\ \ \bar{3}\ .\ 0\ \ \bar{2}\ \ \bar{1}\qquad}$$

The two roots are therefore given approximately by $x = 0.16375$ and 6.979

We could have used other transformations instead of the one used here, to get the second root—we could, for example, have reduced the roots by 7.

Also since the sum of the roots of a quadratic equation is minus the coefficient of x divided by the coefficient of x^2, the second root can be easily obtained from the first.

Example 12   Solve $x^2 - 4x - 6 = 0$

Here both the roots are numerically greater than 1, and we may find, either by observation or by using the formula—the first differential is the square root of the discrimminant (see 'Vedic Mathematics' page 158)—that there is a root at about $x = 5$.

We may therefore reduce the roots of this equation by 5, as shown in Chapter 8.

☐ $h = -5$ and if $Q(x) = x^2 - 4x - 6$, then $Q(5) = -1$
$\qquad\qquad Q'(x) = 2x - 4,\qquad\qquad Q'(5) = 6$

So the transformed equation is $y^2 + 6y - 1 = 0$

☐ $y = 0.1\,6 - \dfrac{y^2}{6}$

$$
\begin{array}{c|ccccccc}
\frac{1}{6} & 0 & . & 2 & \bar{3} & \bar{3} & \bar{3} & \bar{3} \\
-\frac{y^2}{6} & & & & 1\bar{2} & 5\bar{6} & 6\,0 \\
\hline
y = 0 & . & 2 & \bar{4} & 2 & 3
\end{array}
$$

Therefore y = 0.1623 is one root of this transformed equation, so that one root of the original equation is x = 5.1623. And since the sum of the roots of the original equation is 4, the other root must be x = –1.1623.

Chapter 11

TRANSCENDENTAL EQUATIONS

The methods described in the last chapter are also very suitable for the solution of transcendental equations.

Example 1   Solve the transcendental equation $x + \sin x = 1$

Using the series expansion for sinx: $x + x - \frac{x^3}{3!} + \frac{x^5}{5!} - \ldots = 1$

$\square$ $x = 0.5 + \frac{x^3}{12} - \frac{x^5}{240} + \frac{x^7}{2\%7!} - \ldots$

$\therefore$

		0 . 5						
$\frac{x^3}{12}$	12			1_3	1_1	1	2	$2_{\bar{2}}$
$-\frac{x^5}{240}$	2					$\bar{1}_1$	$\bar{5}_i$	
x =		0 . 5	1	1	0	3		
(x² =		. 3 $_5$	$\bar{4}_0$	1_1	1_2	$\bar{1}_1$	)	

The method, which follows similar steps to those from the previous chapter, is quick and efficient if x is small and can be extended to any number of decimal places.

Using the figures of the answer to get the next answer figure, the procedure is similar to that in iterative techniques. But the beauty of this method is that only those figures actually required to get the next figure are calculated so that not a single superfluous step is taken: the method is one hundred per cent efficient. And with the extra flexibility offered by the vinculum device the figures can always be kept small and manageable.

The next example shows the solution of Kepler's Equation, an equation of great importance in positional astronomy. This is
$$M = E - e \sin E$$

where M and E are the mean and eccentric anomalies respectively of a planet in its orbit and e is the eccentricity of the orbit.

The eccentricity, e, is a measure of the oblate ness of the ellipse and can be defined by $e^2 = 1 - \frac{b^2}{a^2}$ where a and b are the lengths of the semi-major and semi-minor axes:

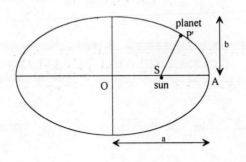

The velocity of a planet in an elliptical orbit is variable and the mean anomaly, M, is the angle shown in the diagram opposite, assuming the planet travelled at a constant speed that completed the orbit in the same time as the actual planet and started at A at the same instant.

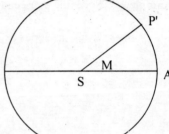

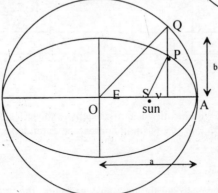

The eccentric anomaly, E, is as shown in the diagram above where Q lies on a circle of radius a. v is the true anomaly of the planet at P and there is a fairly simple formula for getting v once E is known.

So the problem is to find E given M and e in the equation $M = E - e\sin E$

We will take $M = 0.30303$ radians and $e = 0.016722$, which is the eccentricity of the Earth's orbit.

Example 2 Solve $0.30303 = E - 0.016722\sin E$

$$E = 0.30303 + 0.016722\left(E - \tfrac{E^3}{3!} + \tfrac{E^5}{5!} - \ldots\right)$$

M	0 . 3	0	3	0	3	0	0		1
eE		1 $_{\bar4}$	$\bar5$ $_3$	2 $_{\bar6}$	$\bar5$ $_1$	2 $_1$	1 $_{\bar3}$		2
$-\tfrac{eE^3}{3!}$ 6			$\bar1$ $_i$	2 $_i$	$\bar1$ $_3$	$\bar5$ $_0$			3
$\tfrac{eE^5}{5!}$ 2						4 $_0$			4
E =	0 . 3	1	$\bar2$	1	0	1	0		5
(E² =	. 1 $_i$	0 $_{\bar4}$	$\bar5$ $_i$	$\bar1$	2	3 $_{\bar4}$	)		6

where $e = 0.016722 = 0.02\bar3\bar322$

Bring 0.3 into the answer row.

Row 2: we multiply e and E, $CP\begin{pmatrix}2\\3\end{pmatrix} = 6 = 1_{\bar4}$

Bring 1 into the answer row.

Row 2: $CP\begin{pmatrix}2 & \bar3\\3 & 1\end{pmatrix} = \bar7,\ \bar7 + \overline{40} = \bar5_3$

Bring $\bar2$ down into the answer row.

Row 2: $CP\begin{pmatrix}2 & \bar3 & \bar3\\3 & 1 & \bar2\end{pmatrix} = \overline{16},\ \overline{16} + 30 = 2_{\bar6}$

Row 3: we find E² (row 6) then multiply rows 2 and 6, divide by 6 and change the sign: $CP\begin{pmatrix}1\\1\end{pmatrix} = 1,\ CP\begin{pmatrix}1 & \bar5\\1 & 0\end{pmatrix} = \bar5$, giving $1\bar5 = 5$, $5 \div 6 = 1\ r\bar1$, put $\bar1_i$

Bring 1 down into the answer.
Etc.

We get $\underline{E = 0.3081010}$ to 7 decimal places.

Although M, and therefore E, take all values between 0 and 2π, by using the angle to the nearest quadrant boundary we can always arrange that the quantity we calculate is less than about $\tfrac{\pi}{4}$. For example, if M=3 radians, so that $E \approx 3$, we let $E = \pi - A$:

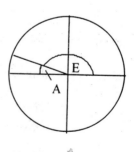

$$\therefore M = \pi - A - e\sin(\pi - A)$$
$$= \pi - A - e\sin A$$

from which we can find A and then E, since $E = \pi - A$.

Transcendental equations occur in a wide range of situations including the prediction of eclipses.

Example 3   Solve $x + e^x = 2$

Using the expansion for e^x we get

$$x + 1 + x + \tfrac{x^2}{2!} + \tfrac{x^3}{3!} + \tfrac{x^4}{4!} + \tfrac{x^5}{5!} + \ldots = 2$$

$$\therefore x = 0.5 - \tfrac{x^2}{4} - \tfrac{x^3}{12} - \tfrac{x^4}{48} - \tfrac{x^5}{240} - \ldots$$

0.5	0 . 5							1
$-\tfrac{x^2}{4}$ 4		$\bar{5}$	5	$\overset{+}{1}$	$\underset{6}{}$	0	6	2
$-\tfrac{x^3}{12}$ 3		$\bar{1}$		3	$\underset{4}{}$	$\bar{2}$	1	3
$-\tfrac{x^4}{48}$ 4		$\bar{1}$	$_i$			2	3	4
$-\tfrac{x^5}{240}$ 5						$\bar{1}$	0	5
x =	0 . 5	$\bar{6}$		3		$\bar{1}$		6

Bring down 0.5 to row 6.
Row 2: D(5) = 25, 25 ÷ 4 = 5 r5, put $\bar{5}_5$

Row 3: we multiply rows 2 and 6 and divide by 3, $CP\begin{pmatrix}5\\5\end{pmatrix} = \overline{25}$

$$\overline{25} + 3 = \bar{7}\,\bar{r4},\ \text{put } \bar{1}3_{\bar{4}}$$

Bring down $\bar{6}$
Row 2: $D(5\bar{6}) = \overline{60}$, $\overline{60} + 50 = \overline{10}$, $\overline{10} \div 4 = \bar{1}$ r6, put $1_{\bar{6}}$

Row 4: we multiply rows 3 and 6 and divide by 4, $CP\begin{pmatrix}\bar{1}\\5\end{pmatrix} = \bar{5},\ \bar{5} \div 4 = \bar{1}i$

Bring down 3 into the answer.
Etc.

Example 4   Solve $x = \cos x + 1$

This becomes $x = 2 - \tfrac{x^2}{2!} + \tfrac{x^4}{4!} - \tfrac{x^6}{6!} + \ldots$

There is a solution to this equation at about $x = 1.3$, but this would converge rather too slowly so we may try transforming to another equation.

Suppose we choose to construct an equation with roots ¼ of these.
The transformed equation is (see Chapter 8):

$$x = \tfrac{2}{4} - \tfrac{x^2}{2!}\,4 + \tfrac{x^4}{4!}\,4^3 - \tfrac{x^6}{6!}\,4^5 + \ldots$$

i.e. $x = 0.5 - 2x^2 + \tfrac{8}{3}x^4 - \tfrac{64}{45}x^6 + \ldots$

0.5		$0 \; . \; 5$	1
$-2x^2$		$\bar{2} \;_{\bar{2}} \; 0 \;_4 \; \bar{6} \;_0 \; \bar{1} \;^{+}_{\bar{6}}$	2
$\frac{8}{3}x^4$	$\frac{2}{3}$	$2 \;_2 \; 8 \;_{\bar{4}} \; 2 \;_2$	3
$-\frac{64}{45}x^6$	$\frac{4}{15}$	$\bar{1} \;_{\bar{1}} \; \bar{5} \;_1$	4
$x \; =$		$0 \; . \; 3 \quad 2 \quad 1 \quad \bar{2}$	5

So the solution to the original equation is $4 \times 0.321\bar{2} = \underline{1.283}$ to 4 sig. figures.

We do not bring 0.5 straight down into the answer: if we do this the next term would eliminate it. But starting with 0.3 we find that the second term, when combined with row 1, gives 0.3 also: $D(3) = 9$, $9 \times 2 = 2\bar{2}$, so we put down $\bar{2}_{\bar{2}}$

Row 3: we square row 2, multiply by 2 and divide by 3,
$D(\bar{2}) = 4$, $4\%2 = 8$, $8 + 3 = 2_2$

Row 2: we have 0.32 so far in row 5 and this gives us 24 in row 2, which is just what we require as we have $\bar{2}0$ there already. We therefore put down 0_4

Bring down 2 into the answer.

And so on.

Chapter 12

SOLUTION OF CUBIC AND HIGHER ORDER EQUATIONS

Bharati Krishna gives a procedure for extracting cube roots by the Urdhvatiryak sutra in 'Vedic Mathematics' (Chapter 36). This extends naturally to the solution of cubic equations and to higher order equations generally. Much work remains to be done, especially as regards convergence, but an outline of the method, and an illustration of its essential economy, is given by the examples below.

Example 1   Suppose we have to calculate the cube root of 70.1, or, what is the same thing, solve the cubic equation

$$f(x) \equiv x^3 - 70.1 = 0$$

As in the square root procedure (see Chapter 1 above) we represent the answer in the decimal form

$$x = a.bcde \ldots.$$

that is

$$x = a + 10^{-1}b + 10^{-2}c + 10^{-3}d + \ldots.$$

where a, b, c, d .... are decimal digits 1, ...9, or 0.

The cubic equation to be solved can be expressed

$$f(a + 10^{-1}b + 10^{-2}c + \ldots.) = 0$$

or rearranging

$$10^{-1}b + 10^{-2}c + 10^{-3}d + \ldots =$$

$$\{-f(a) - \tfrac{1}{2}f''(a)[10^{-1}b + 10^{-2}c + \ldots]^2 - \tfrac{1}{6}f'''(a)[10^{-1}b + 10^{-2}c + 10^{-3}d + \ldots]^3\} \div f'(a)$$

where f', f" and f''' denote successive differentials of $f(x) \equiv x^3$.

In this form an initial value a (here a = 4) leads to a recurrence relation in which each digit b, c, d . . . is derived from its predecessors by equating appropriate powers of 10. In the mental work which accompanies the pencil-and-paper solution, it is the coefficient of these powers of ten which follow from the Urdhvatiryak sutra.

The working begins with the choice (by inspection or 'Vilokanam') **a = 4**. We then calculate the differential terms $f'(a) = 48$, $\frac{1}{2}f''(a) = 12$, $\frac{1}{6}f'''(a) = 1$, and also the order 1 component of the remainder $-f(a)$, which is 6. The first differential $f'(a)$ becomes a divisor which remains constant throughout the working, while the remainder term carried over to the next column (order 10^{-1}) gives a total remainder 61 which becomes the dividend for the next step. The calculation then proceeds thus:

$$
\begin{array}{c|ccccccccc}
48 & 70 . & _6 & 1 & _{13}0 & _{22}0 & _{27}0 & _{48}0 \\
12 & & & & \overline{12} & \overline{48} & \overline{120} & \overline{192} \\
\underline{1} & & & & & \overline{1} & \overline{6} & \overline{21} \\
\hline
& 4 ./ & 1 & 2 & 3 & 2 & 5
\end{array}
$$

The steps are

(i) Divide the first dividend 61 by the divisor 48, giving **b=1** and leaving a remainder 13 to be carried over to the next column.

(ii) The new dividend (order 10^{-2}) is
130 − 12 [the second differential $\frac{1}{2}f''(a)]\times1$ [b^2] = 118, 118 ÷ 48 gives **c=2**, remainder 22.

(iii) Now the dividend (order 10^{-3}) is
220 − 12 [$\frac{1}{2}f''(a)] \times 4$ [2bc] − 1 [$\frac{1}{6}f'''(a)] \times 1$ [b^3] = 171, 171 ÷ 48 gives **d=3**, remainder 27.

(iv) 270 − 12 × 10 [2bd + c^2] − 1 × 6 [$3b^2c$] = 144, 144 ÷ 48 gives **e=2** if the remainder 48 is to be sufficient to meet the demands of the next step.

(v) 480 − 12 × 16 [2be + 2cd] − 1 × 21 [$3bc^2 + 3b^2d$] = 267,
267 ÷ 48 = **5** remainder 27

and so on.

Notice the use of the sutra in calculating the multipliers of $\frac{1}{2}f''(a)$ as the Dvandvayoga or duplex terms b^2, 2bc, 2bd + c^2, 2be + 2cd, . . . These are the coefficients of successive powers of 10^{-1} in the expansion of
$$(10^{-1}b + 10^{-2}c + 10^{-3}d +...)^2$$

Similarly, the triplex terms b^3, $3b^2c$, $3bc^2 + 3b^2d$, ... which multiply $\frac{1}{6}f'''(a)$ are coefficients in the expansion of

$$(10^{-1}b + 10^{-2}c + 10^{-3}d + ...)^3$$

They in turn follow from the vertically and crosswise multiplication of the decimal digits of the solution b, c, d, e, ... with the duplex sequence b^2, $2bc$, $c^2 + 2bd$, $2be + 2cd$, ...

Example 2   Exactly the same process applies in the solution of
$f(x) \equiv 2x^3 + 5x^2 + 6x - 64 = 0$

The difference in algebraic structure between this cubic equation and the simple root of Example 1 are irrelevant to the direct digit by digit solution.

In this example where again x = a.bcde ...
we have **a = 2**, $f'(a) = 50$, $\frac{1}{2}f''(a) = 17$, $\frac{1}{6}f'''(a) = 2$,
and the initial remainder $-f(a) = 16$.
The working becomes

$$
\begin{array}{c|cccccccc}
50 & 64 & . & {}_{16}0 & {}_{10}0 & {}_{\bar{3}} & 0 & {}_{18}0 \\
17 &   &   &   & \overline{153} & 102 & \overline{17} \\
2 &   &   &   &   & \overline{54} & 54 \\
\hline
   & 2 & . / & 3 & \bar{1} & 0 & 4
\end{array}
$$

where the steps are
(i). $160 \div 50$ gives **b = 3**, remainder 10.
(ii) $100 - 17 \times 3^2 = -53$, $-53 \div 50$ gives $c = \bar{1}$, remainder $\bar{3}$
(iii) $-30 - 17 \times (2 \times 3 \times \bar{1}) - 2 \times 3^3 = 18$, $18 \div 50$ gives **d = 0**, remainder 18
(iv) $180 - 17 \times (2 \times 3 \times 0 + \bar{1}^2) - 2 \times 3^2 \times \bar{1} = 217$,
          $217 \div 50 = 4$ remainder 17, and so on.

In this example the solution $2.3\bar{1}04$ must be converted finally to the positive digit form $2.3\bar{1}04 = 2.2904...$ The use of the vinculum digit in $2.3\bar{1}...$, in preference to 2.29..., leads to much smaller and more manageable coefficients for the correction terms which multiply the higher order derivatives $\frac{1}{2}f''(a)$, $\frac{1}{6}f'''(a)$.

Example 3   Consider the equation given as Example 20a in Chapter 12, Cubic Equations, of 'Introductory Lectures on Vedic Mathematics'.

This is $f(x) = x^3 - 7x^2 + 20x - 37 = 0$

The calculation scheme of Examples 1 and 2 leads to an initial digit **a = 4**, $f'(a) = 12$, $\frac{1}{2}f''(a) = 5$, $\frac{1}{6}f'''(a) = 1$, and the first few steps

$$
\begin{array}{c|cccccc}
12 & 3 & 7 \,.\, {}_5 0 \, {}_2 0 \\
5 & & & 80 \\
1 & \\
\hline
& & 4 \,.\, /\ 4 & \overline{5}
\end{array}
$$

The working is already awkward because the duplex and triplex corrections become large compared with the divisor 12.

One device to improve the convergence is to work with the transformed equation.

$$g(z) = z^3 - 2 \times 7z^2 + 2^2 \times 20z - 2^3 \times 37 = 0 \quad \text{(see Chapter 8)}$$

or $\quad g(z) = z^3 - 14z^2 + 80z - 296 = 0$

where $z = 2x$.

Now the first digit of z is 9, the differential terms are $g'(9) = 71$, $\tfrac{1}{2}g''(9) = 13$, $\tfrac{1}{6}g'''(9) = 1$, and the first remainder is $-g(9) = \overline{19}$

Now the solution becomes

$$
\begin{array}{c|cccccccc}
71 & 296 \,.\, & {}_{\overline{19}} 0 & {}_{23} 0 & {}_{\overline{29}} 0 & {}_{35} 0 \\
13 & & & \overline{117} & 156 & \overline{208} \\
1 & & & & 27 & 54 \\
\hline
& & 9 \,.\, / & \overline{3} & 2 & \overline{2} & 1
\end{array}
$$

so $z = 9.\overline{3}2\overline{2}1 = 8.7181$, $x = 4.3590 \ldots$

Example 4 Solve $\quad f(x) = x^5 + 3x^4 + 5x^3 + 7x^2 + 19x - 51 = 0$

The method of solution applied above to cubic equations generalises directly to this fifth order equation.
With the solution x = a.bcde... the equation may be expressed in the form

$$10^{-1}b + 10^{-2}c + 10^{-3}d + \ldots = \{-f(a) - \tfrac{1}{2}f''(a)[10^{-1}b + 10^{-2}c + 10^{-3}d + \ldots]^2$$

$$-\tfrac{1}{6}f'''(a)[10^{-1}b + 10^{-2}c + \ldots]^3$$

$$-\tfrac{1}{24}f^{iv}(a)[10^{-1}b + 10^{-2}c + \ldots]^4$$

$$-\tfrac{1}{120}f^{v}(a)[10^{-1}b + 10^{-2}c + \ldots]^5\} \div f'(a)$$

which again gives a recurrence relation giving the desired digits b, c, d . . . from its predecessors.

To find the solution near $x = 1$ the first step is to set $\mathbf{a} = \mathbf{1}$ and then calculate the successive differentials

$$f'(a) \quad = \quad 5a^4 + 12a^3 + 15a^2 + 14a + 19 = 65$$

$$\tfrac{1}{2}f''(a) = \quad \tfrac{1}{2}\,[20a^3 + 36a^2 + 30a + 14] = 50$$

$$\tfrac{1}{6}f'''(a) = \quad \tfrac{1}{6}\,[60a^2 + 72a + 30] = 27$$

$$\tfrac{1}{24}f^{iv}(a) = \quad \tfrac{1}{24}\,[120a + 72] = 8$$

$$\tfrac{1}{120}f^{v}(a) = \quad \tfrac{1}{120}\,[120] = 1$$

The calculation then follows thus

$$
\begin{array}{c|cccccccc}
65 & 51\;.\;{}_{16}\,0\;\;{}_{30}\,0\;\;{}_{35}\,0\;\;{}_{i}\;\;0 \\
50 & \qquad\qquad\quad \overline{200}\;\;\overline{200}\;\;150 \\
27 & \qquad\qquad\qquad\qquad\;\;\; \overline{216}\;\;\overline{324} \\
8 & \qquad\qquad\qquad\qquad\qquad\qquad \overline{128} \\
1 \\
\hline
& \quad 1\;.\;/\;2\quad\;\; 1\quad\;\; \overline{1}\quad\;\; \overline{4}
\end{array}
$$

The individual steps are

(i)   The first remainder $-f(1)$ is 16, which becomes 160×10^{-1}, and
      $160 \div 65$ gives $\mathbf{b = 2}$, remainder 30

(ii) $300 - 50\,[\tfrac{1}{2}f''(1)] \times 4\,[b^2] = 100$, $100 \div 65$ gives $\mathbf{c = 1}$, remainder 35

(iii) $350 - 50 \times 4\,[2bc] - 27\,[\tfrac{1}{6}f'''(1)] \times 8\,[b^3] = -66$,
      $-66 \div 65$ gives $\mathbf{d = 1}$, remainder -1

(iv) $-10 - 50 \times \overline{3}\,[2bd + c^2] - 27 \times 12\,[3b^2c] - 8\,[\tfrac{1}{24}f^{iv}(1)] \times 16\,[b^4] = -312$,
      $-312 \div 65$ gives $\mathbf{c = -4}$, remainder -52

and so on. Further decimal digits will become harder to calculate with the proliferation of the subtraction terms but the first four decimal places

$$x = 1.21\overline{1}\overline{4} = 1.2086...$$

are found very quickly and economically.

Example 5 The reader may like to follow through the further generalisation needed to extract the (exact) root of the seventh degree equation

$$f(x) = x^7 + x^5 + 3x^3 - 248.7328641 = 0$$

With **a** = **2** the initial remainder (order 1) is 64, and the successive differential terms in what is in effect the Taylor expansion of $f(x)$ are $f'(a) = 564$, $\frac{1}{2}f''(a) = 770$, $\frac{1}{6}f'''(a) = 603$, $\frac{1}{24}f^{iv}(a) = 290$, $\frac{1}{120}f^{v}(a) = 85$, $\frac{1}{720}f^{vi}(a) = 14$, $\frac{1}{5040}f^{vii}(a) = 1$.

The solution is here greatly simplified in as much as

$$x = a.b$$

with c = 0, d = 0, e = 0, . . . etc.,

but setting out the calculation in the same way as the other examples we would find:

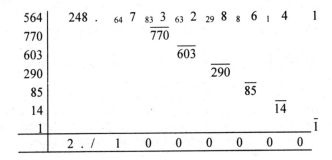

All the other subtraction terms are identically zero, and we find the exact solution x = 2.1.

Chapter 13

FUNCTIONS OF POLYNOMIALS

A well-known method of obtaining a power series for any function makes direct use of Taylor's series or Maclaurin's series. The drawback is that higher order derivatives tend to become increasingly difficult to evaluate and an approach free from such a disadvantage would be welcome.

The concern of this chapter is to outline procedures for finding power series expansions of various functions of polynomials. Thus, if $P(x)$ and $Q(x)$ are polynomials in x, the interest is in ways of deriving expansions for $P(x) \times Q(x)$, $P^2(x)$, $P(x) \div Q(x)$, $\dfrac{1}{P(x)}$, $\sqrt{P(x)}$, $\sqrt[3]{P(x)}$, $\sqrt[5]{P(x)}$, $P^m(x)$ (where m is a constant, integral or fractional), $\ln P(x)$, $\exp P(x)$ and $\cos P(x)$.

Amongst applications are the solution of transcendental equations, and the solution of differential equations—especially non-linear ones. One or two simple examples of these two applications are given in Chapter 15. That more can be done with this method is shown by solving an integral equation, and then an integro-differential equation.

Of the above-mentioned functions, the procedures for $P(x).Q(x)$ and $P^2(x)$ have already been given in Chapter 1, and for $\ln P(x)$ and $\exp P(x)$ in Chapter 7.

We therefore begin by considering division.

DIVISION

Example 1 $(2 + 3x + 5x^2 + 3x^3) \div (2 - x + 3x^2)$

Here we will use the algebraic equivalent of the 'vertically and crosswise' method of division explained in Chapter 1. The complete working and solution is written down as follows:

$$
\begin{array}{c|ccccccc}
-1 \quad 3 & 2 & + \, 3x & + \, 5x^2 & + \, 3x^3 \\
2 & \\
\text{Solution} & 1 & + \, 2x & + \, 2x^2 & - \frac{1}{2}x^3 & - \, 3\frac{1}{4}x^4 & - \frac{7}{8}x^5 & + \frac{5}{48}x^6 + \dots
\end{array}
$$

The steps are as follows:

(i) $2 \div 2 = 1$ (Write it down)

(ii) We now multiply this 1 by the first flag digit (-1), subtract from 3 (the coefficient of x), and divide the result by 2:

$$\{3 - [1 \times (-1)]\} \div 2 = 2$$

This is written down, and appears finally as the coefficient of x.
But at this stage powers of x can conveniently be omitted from the solution.

So far we have:

$$
\begin{array}{c|cccc}
-1 \quad 3 & 2 & + \, 3x & + \, 5x^2 & + \, 3x^3 \\
2 & \\
\hline
& 1 & 2
\end{array}
$$

(iii) The two solution coefficients, 1 and 2, are now multiplied crosswise by the two flag digits:
$$(-1)(2) + 1 \times 3 = 1$$

Taking the result from 5 (the coefficient of x^2), and dividing by 2, we have:

$$(5 - 1) \div 2 = 2$$

This is the next coefficient in the solution.

(iv) The two most recent solution coefficients (2 and 2) are now multiplied crosswise by the two flag digits, giving:

$$(-1)(2) + 2 \times 3 = 4$$

Taking this 4 from 3 (the coefficient of x^3), and dividing by 2, we have the next solution term:
$$(3 - 4) \div 2 = -\tfrac{1}{2}$$

And so on.

If we chose to stop dividing after three terms the answer could be written:

$$
\begin{array}{cc|cccc}
-1 & 3 & 2 + 3x & + 5x^2 & + 3x^3 & \\
2 & & & & & \\
\hline
& & 1 & 2x & + 2x^2 & + \dfrac{-x^3-6x^4}{2-x+3x^2}
\end{array}
$$

The subtractions involving flag-digits are now truncated, since remainder terms do not participate in such subtractions.

The last two terms are obtained from: $3 - [(-1)(2) + 3(2)] = -1$
and $0 - 2\times3$ $= -6$

A method of division yielding highest powers first and then successively lower powers is given in 'Vedic Mathematics'.

RECIPROCALS

These are a special case of division. We might reasonably expect this to be a rewarding area of study to the Vedic mathematician, special cases having an honoured place in his scheme. Here, however, we shall simply content ourselves with an example.

Example 2   Evaluate $\dfrac{1}{1-x-x^2}$ as a power series in x.

Working and solution:

$$
\begin{array}{cc|l}
-1 & -1 & 1 \\
1 & & \\
\hline
\text{Solution} & & 1 + x + 2x^2 + 3x^3 + 5x^4 + 8x^5 + 13x^6 + 21x^7 + \ldots
\end{array}
$$

SQUARE ROOTS

Example 3   Obtain $\sqrt{4+12x+13x^2+6x^3+x^4}$ as a power series in x.

Working and solution:

$$
\begin{array}{c|ccccc}
4 & 4 & + 12x & + 13x^2 & + 6x^3 & + x^4 \\
\hline
\text{Answer:} & \pm 2 /+ & 3x & + 1x^2 & + 0x^3 & + 0x^4
\end{array}
$$

Check: $(2 + 3x + x^2)^2 = 4 + 12x + 13x^2 + 6x^3 + x^4$

The procedure is the same as for determining the square root of a number, except that there is no carrying. The steps are:

(i) $\sqrt{4} = 2$

This leads to one solution, and $-\sqrt{4} = -2$ leads to the second. The **2** is written down as the first term of the answer.

(ii)   Doubling this 2, we obtain the divisor, 4. A diagonal stroke now separate this 2 from those ensuing terms which contribute to the duplex (viz 3^2, $2\times3\times1$, $2\times3\times0 + 1^2$, $2\times3\times0 + 2\times1\times0$ etc.)

(iii) The second term in the answer is given by $12 \div 4 = $ **3**.

(iv) $13 - 3^2 = 4$
and $4 \div 4 = $ **1**, which is the third coefficient of the answer.

(v)   The duplex is now $2\times3\times1 = 6$, and $6 - 6 = 0$,
so that $0 \div 4 = $ **0** (fourth coefficient of the solution)

(vi)   The duplex is $2\times3\times0 + 1^2 = 1$, and $1 - 1 = 0$,
whence $0 \div 4 = $ **0** (fifth coefficient of the solution)

All subsequent terms are found to be zero, and so the solution is exact.

Example 4   Obtain $\sqrt{4 + 12x + 13x^2 + 7x^3 + 2x^4}$ as a power series in x.

The first three terms being the same as in the previous example, the working starts off the same. The working and solution proceeds thus:

$$
\begin{array}{r|l}
4 & 4 + 12x + 13x^2 + 8x^3 + 4x^4 \\
\hline
\text{Solution:} & \pm (2/+ \quad 3x + x^2 + \tfrac{1}{2}x^3 + 0x^4 - \tfrac{1}{4}x^5 + \tfrac{5}{16}x^6 + \ldots)
\end{array}
$$

The procedure for obtaining the square root in descending powers of x is the same, except that the coefficients are considered in the reverse order.

Example 5   Obtain $\sqrt{4x^4 + 8x^3 + 13x^2 + 12x + 4}$ as a descending power series in x.

The solution proceeds thus:

$$
\begin{array}{r|l}
4 & 4x^4 + 8x^3 + 13x^2 + 12x + 4 \\
\hline
& 2x^2 / + 2x + \tfrac{9}{4} + \tfrac{3}{4x} - \tfrac{65}{64x^2} + \tfrac{11}{64x^3} + \ldots
\end{array}
$$

In general the square root of a polynomial leads to an infinite series.

CUBE ROOT

The method is essentially the same as that for finding the cube root, etc., of a number, for which the basis is outlined in Chapter 12. The difference with a polynomial is that there is no carrying.

Only a very brief explanation of the procedure is outlined here, starting with an example which works out exactly, for simplicity of demonstration.

Example 6   Find $\sqrt[3]{x^6 + 12x^5 + 57x^4 + 136x^3 + 171x^2 + 108x + 27}$ obtaining the solution as a descending power series in x.

The complete working and solution is written down as follows:

$$
\begin{array}{c|ccccccc}
3 & x^6 & + & 12x^5 & + & 57x^4 & + & 136x^3 & + & 171x^2 & + & 108x & + & 27 \\
\hline
\text{Solution:} & 1x^2 / & + & 4x & + & 3 & + & \frac{0}{x} & + & \frac{0}{x^2} & + & \frac{0}{x^3} + & \frac{0}{x^4}
\end{array}
$$

Steps:

(i)   Coefficient of x^2: write down **1** (since $1 - 1^3 = 0$)

(ii)   The divisor is thrice this 1

(iii)   For the coefficient of x: $12 \div 3 = \textbf{4}$ remainder 0

(iv) $57 - 3 \times 1 \times 4^2 = 9$

(v)   For the coefficient of unity: $9 \div 3 = \textbf{3}$ remainder 0

(vi) $6 \times 1 \times 4 \times 3 + 4^3 = 72 + 64 = 136$

(vii) $136 - 136 = \textbf{0}$

(viii) For the coefficient of $\frac{1}{x}$: $0 \div 3 = \textbf{0}$ remainder 0

(ix) $6 \times 1 \times 4 \times 0 + 3 \times 1 \times 3^2 + 3 \times 4^2 \times 3 = 171$

(x) $171 - 171 = 0$

(xi)   For the coefficient of $\frac{1}{x^2}$: $0 \div 3 = \textbf{0}$ remainder 0

(xii) $6 \times 1 \times 4 \times 0 + 6 \times 1 \times 3 \times 0 + 3 \times 4 \times 3^2 = 108$

(xiii) $108 - 108 = 0$

(xiv)   For the coefficient of $\frac{1}{x^3}$: $0 \div 3 = \textbf{0}$ remainder 0

(xv) $3^3 = 27$

(xvi) $27 - 27 = 0$

(xvii) For the coefficient of $\frac{1}{x^4}$: $0 \div 3 = \mathbf{0}$ remainder 0

$\therefore$ $\underline{x^2 + 4x + 3 \text{ is an exact cube root}}$

All three cube roots are obtained by multiplying by the cube roots of unity, giving $\frac{\sqrt{3}}{4}(-1 \pm 2i)(x^2 + 4x + 3)$ as the other solutions, where $i = \sqrt{-1}$

The justification of the procedure is contained in a rearrangement of the terms of the multinomial:
$$(a + bx + cx^2 + \ldots)^3 = a^3 + 3a^2(bx + cx^2 + dx^3 + \ldots) + 3a(bx + cx^2 + dx^3 + \ldots)^2$$
$$+ (bx + cx^2 + dx^3 + \ldots)^3$$

Example 7   Obtain $\sqrt[3]{27 + 108x + 198x^2 + 208x^3 + 105x^4 - 78x^5}$ as an ascending power series in x.

The working and first steps of the solution can be written down as follows:

$$3a^2 = 27 \overline{\left| 27 + 108x + 198x^2 + 208x^3 + 105x^4 - 78x^5 \right.}$$

Solution	3 +	4x +	$2x^2$ +	$0x^3$ −	$1x^4$ -	$2x^5$ + ...
$3a = 9$			16	16	4	-8 (duplex terms)
				64	96	48 (triplex terms)

FIFTH ROOT

Example 8   The reader may care to use what is essentially the same process as has just been outlined to verify that $(2 + x)$ is one root of:

$$(32 + 80x + 80x^2 + 40x^3 + 10x^4 + x^5)^{\frac{1}{5}}$$

The procedure rests on a rearrangement of the terms of the multinomial:

$$(a + bx + cx^2 + \ldots)^5 = a^5 + 5a^4(bx + cx^2 + \ldots) + 10a^3 (bx + cx^2 + \ldots)^2$$
$$+ 10a^2 (bx + cx^2 + \ldots)^3 + 5a (bx + cx^2 + \ldots)^4$$
$$+ (bx + cx^2 + \ldots)^5$$

The leading term is now $a = 2$, and the divisor is $5a^4$
The working and solution can be written down as follows:

$$5a^4 = 80 \quad \left| \quad \frac{32 + 80x + 80x^2 + 40x^3 + 10x^4 + x^5}{2 \,/\, 1 \quad 0 \quad\quad 0 \quad\quad 0 \quad\quad 0} \right.$$

$10a^3 = 80$			1	0	0	0
$10a^2 = 40$				1	0	0
$5a = 10$					1	0
1						1

Answer <u>2 + x</u>

All five roots are given by multiplying by the fifth roots of unity, i.e. 1, $\exp(i2\pi/5)$, $\exp(i4\pi/5)$, $\exp(i6\pi/5)$, $\exp(i8\pi/5)$.

POWERS

Here we arc concerned with evaluation of $P^m(x)$, where 'm' is a constant.
We start with a simple example which can readily be checked.

Example 9   Expand $(1 + x)^4$

Let $(1 + x)^4 = A + Bx + Cx^2 + Dx^3 +$ ...(1)

Take logs and differentiate with respect to x, and then equate coefficients.

$$\frac{4}{1+x} = \frac{\overset{4}{B} + \overset{12}{2Cx} + \overset{12}{3Dx^2} + \overset{4}{4Ex^3} + \overset{0}{5Fx^4} + ...}{\underset{1}{A} + \underset{4}{Bx} + \underset{6}{Cx^2} + \underset{4}{Dx^3} + \underset{1}{Ex^4} + ...} \qquad ...(2)$$

The first coefficient, A, is obtained by putting $x = 0$ in Equation (1), and subsequent coefficients come from Equation (2).

$$\therefore (1 + x)^4 = \underline{1 + 4x + 6x^2 + 4x^3 + x^4}$$

This is recognisable as the relevant row of Pascal's triangle.

Example 10   Obtain a series expansion for $(2 + x + 2x^2)^3$.

Take logs, differentiate with respect to x, and equate coefficients, to obtain:

$$\frac{3(1+4x)}{2+x+x^2} = \cdot \frac{\overset{12}{B} + \overset{60}{2Cx} + \overset{75}{3Dx^2} + \overset{120}{4Ex^3} + 5Fx^4 + ...}{\underset{8}{A} + \underset{12}{Bx} + \underset{30}{Cx^2} + \underset{25}{Dx^3} + \underset{30}{Ex^4} + ...}$$

The symmetry halves the work in this example, and also acts as a check.

$$\therefore (2 + x + 2x^2)^3 = \underline{8 + 12x + 30x^2 + 25x^2 + 30x^4 + 12x^5 + 8x^6}$$

Example 11 Expand $(1 + 3x - 2x^2)^{10}$

Let $(1 + 3x - 2x^2)^{10} = a + bx + cx^2 + dx^3 + \ldots$

Take logs, differentiate with respect to 'x', and equate coefficients, to obtain:

$$\frac{10(3-4x)}{1+3x-x^2} = \frac{\overset{30}{B} + \overset{770}{2Cx} + \overset{8100}{3Dx^2} + \overset{42840}{4Ex^3} + 5Fx^4+\ldots}{\underset{1}{A} + \underset{30}{Bx} + \underset{385}{Cx^2} + \underset{2700}{Dx^3} + \underset{10710\ldots}{Ex^4} + Fx^5\ldots}$$

The coefficients, a, b, c, etc., all take integral values, and this acts as a check.

$$\therefore (1 + 3x - 2x^2)^{10} = \underline{1 + 30x + 385x^2 + 2700x^3 + 10710x^4 + \ldots}$$

Putting x = 0.1 in the above expansion gives:

$$1.0302^{-10} = 1.0298^{10} = 1.3413091$$

$$(1 + 0.3 + 0.0385 + 0.002710 + 0.00010711 = \ldots\ldots\ldots = 1.34130810$$

This procedure can be used to find fractional powers of polynomials also.

Example 12 Evaluate the roots of $(1 + 3x - x^2)^{1.3}$ as a power series in x.

Assume that $(1 + 3x - x^2)^{1.3} = a + bx + cx^2 + \ldots$
Take logs, differentiate, and equate coefficients.

$$\frac{1.3(3-2x)}{1+3x-x^2} = \frac{\overset{3.9}{b} + \overset{0.91}{2cx} + \overset{-7.1955}{3dx^2} + 4ex^3 + 5fx^4 +\ldots}{\underset{1}{a} + \underset{3.9}{bx} + \underset{0.455}{cx^2} + \underset{-2.3985}{dx^3} + ex^4 + fx^5 +\ldots}$$

Hence, a solution is:

$$(1 + 3x - x^2)^{1.3} = 1 + 3.9x + 0.455x^2 - 2.3985x^3 + \ldots$$

Putting x = 0.1, e.g. this gives: $\left(1.3\overline{1}\right)^{1.3} = 1.39215.$

The above is one solution. All ten solutions are given by:

$$\left[\exp\left(2i\pi\times0.1n\right)\right](1 + 3.9x + 0.455x^2 - 2.3985x^3 + \ldots) \text{ where } n = 0, 1, 2, \ldots 9$$

These ten solutions arise because the appropriate roots of unity are given by $\exp\left(2i\pi\times1.3n\right)$, where $n = 0, 1, 2, 3 \ldots$, and since the last digit of the three times table goes through the digits 3, 6, 9, 2, 5, 8, 1, 4, 7 and then repeats, the same applies to the 1.3 times table. Integral multiples of $2i\pi$ correspond to so many multiples of unity, and can therefore be ignored.

NATURAL LOGARITHMS AND EXPONENTIALS

The procedure was outlined in Chapter 7, and need not be repeated here. The examples of Chapter 7 were restricted to dealing with real exponents. By considering imaginary exponents, sines and cosines of $P(x)$ are brought into the picture, and this case is considered next.

COSINE AND SINE OF P(x)

Here De Moivres Theorem, $\exp(i\theta) = \cos\theta + i\sin\theta$, will now be combined with the procedure for evaluating $\exp\{P(x)\}$ as a power series, to obtain a general method for obtaining sines and cosines of polynomials.

Example 13　Find a series expansion for $\cos\left(-x + 2x^2 + x^3\right)$

Let
$$\exp i(-x + 2x^2 + x^3) = a + bx + cx^2 + dx^3 + ex^4 + \ldots + i(A + Bx + Cx^2 + Dx^3 + Ex^4 + \ldots) \quad \ldots(3)$$

Differentiating, dividing by Equation (3) and equating coefficients we have:

$$i(-1 + 4x + 3x^2) =$$

$$\begin{array}{cccccccccc}
0 & -1 & 6 & -\frac{23}{6} & -\frac{35}{3} & & -1 & 4 & \frac{7}{2} & -4 & 7\frac{11}{24} \\
\end{array}$$
$$\frac{b+2cx+3dx^2+4ex^3+5fx^4+6gx^5+\ldots+i(B+2Cx+3Dx^2+4Ex^3+5Fx^4+6Gx^5+\ldots)}{a+bx+cx^2+dx^3+ex^4+fx^5+\ldots+i(A+Bx+Cx^2+Dx^3+Ex^4+Fx^5+\ldots)} \quad \ldots(4)$$
$$\begin{array}{cccccccccc}
1 & 0 & -\frac{1}{2} & 2 & -\frac{23}{24} & -\frac{7}{3} & & 0 & -1 & 2 & \frac{7}{6} & -1 & 1\frac{59}{120} \\
\end{array}$$

Here the real and imaginary parts of each coefficient are obtained separately. Thus we first obtain a=1 and A=0 from Equation (3) (on putting x=0).
Thereafter, working with Equation(4) (mentally cross-multiplying) we obtain:

(i)　　b=0　　by considering the coefficient of unity,
　　　　& B=−1 by considering the coefficient of 'i'

(ii) $2c=-1$ by considering the coefficient of x,
& $2C=4$ by considering the coefficient of ix,

(iii) $3d=6$ by considering the coefficient of x^2,
& $3D=\frac{7}{2}$ by considering the coefficient of ix^2,

and so on.

E.g., when equating coefficients of x, we have:

$$i(-1 + 4x)i(0 - x) - (2c + i2C)x$$

$$\therefore \quad i^2[(-1) \times (-1) + 4 \times 0] = 2c$$

The solutions appear in the denominator of Equation (4):

$$\cos(-x + 2x^2 + x^3) = 1 + 0x - \tfrac{1}{2}x^2 + 2x^3 - \tfrac{23}{24}x^4 - \tfrac{7}{3}x^5 + \dots \qquad \dots(5)$$

$$\& \ \sin(-x + 2x^2 + x^3) = -x + 2x^2 + \tfrac{7}{6}x^3 - x^4 + \tfrac{179}{120}x^5 + \dots \qquad \dots(6)$$

We can check by squaring and adding Equations (5) and (6):

$$\cos^2(-x + 2x^2 + x^3) + \sin^2(-x + 2x^2 + x^3) = 1 + 0x + 0x^2 + 0x^3 + 0x^4 + 0x^5 + \dots$$

This verifies that the solutions are correct, the total being one.

Exercise

Evaluate the following:

1) $(4x^3 + 2x^2 + 3x + 2) \div (x^2 - 2x + 1)$

2) $(8x^4 + 3x^2 + 4x + 12) \div (2x^2 - 4x + 6)$

3) $(5x^5 + 2x^4 + 6x^3 + 2x^2 + x + 3) \div (3x^2 + 7x + 1)$

4) $(4x^5 + 3x^4 - 2x^3 + 4x^2 - 2x + 7) \div (4x^2 + 2x + 7)$

5) $\dfrac{1}{1-4x+4x^2}$

6) $\sqrt{x^4 + 4x^3 + 10x^2 + 12x + 9}$

7) $\sqrt{x^4 + 6x^3 + 8x^2 + 9x + 3}$

8) $\sqrt{x^6 + 2x^5 + 6x^4 + 5x^3 + 2x^2 + x - 2}$

9) $\sqrt{x^6 + 3x^5 + 4x^4 + 2x^3 + x^2 + 5x + 1}$

10) $\sqrt[3]{x^3 + 12x^2 + 48x + 64}$

11) $\sqrt[3]{x^6 + 6x^5 + 3x^4 - 28x^3 - 9x^2 + 54x - 27}$

12) $\sqrt[3]{x^9 + 6x^8 + 3x^7 - 25x^6 + 3x^5 + 50x^4 - 60x^3 + 33x^2 - 9x + 1}$

13) $\sqrt[5]{x^5 + 5x^4 + 10x^3 + 10x^2 + 5x + 1}$

14) $\sqrt[5]{x^5 - 10x^4 + 40x^3 - 80x^2 + 80x - 32}$

15) $\sqrt[5]{x^{10} + 15x^9 + 80x^8 + 170x^7 - 35x^6 - 517x^5 - 190x^4 + 600x^3 - 640x^2 + 240x - 32}$

16) $\cos(2x^2 + 3x)$ 17) $\cos(-8x^3 + 4x^2 + 6x)$

18) $\sin(x^3 + 2x^2 + x)$ 19) $\sin(-2x^5 + 4x^3 + 3x^2)$

20) $\operatorname{expi}(3x^3 + 4x^2 + 9x)$

Answers:

1) $2 + 7x + 14x^2 + 25x^3 + \ldots$

2) $2 + 2x + \frac{7}{6}x^2 + \frac{1}{9}x^3 + \frac{55}{54}x^4\ldots$

3) $3 - 20x + 133x^2 - 865x^3 + 5658x^4 - 37006x^5 + \ldots$

4) $1 - \frac{4}{7}x + \frac{8}{49}x^2 - \frac{2}{343}x^3 + \frac{809}{2301}x^4 + \frac{7642}{16107}x^5 + \ldots$

5) $1 + 4x + 12x^2 + 32x^3 + 80x^4 + \ldots$

6) $x^2 + 2x + 3$

   (Note that $(x^4 + 4x^3 + 10x^2 + 12x + 9)^{\frac{1}{2}}$ has two solutions, namely $\ !\ (x^2 + 2x + 3)$)

7) $x^2 + 3x - \frac{1}{2} + \frac{6}{x} - \frac{133}{8x^2}$

8) $x^3 + x^2 + \frac{5x}{2} - \frac{17}{8x} + \frac{21}{8x^2} + \frac{27}{16x^3} + \ldots$

9) $x^3 + \frac{3x^2}{2} + \frac{7x}{8} - \frac{5}{16} + \frac{75}{128x} + \frac{485}{256x^2} - \frac{2973}{512x^3} + \ldots$

10) $x + 4$ 11) $x^2 + 2x - 3$

12) $x^3 + 2x^2 - 3x + 1$ 13) $x + 1$

14) $x - 2$ 15) $x^2 + 3x - 2$

16) $1 - \frac{9x^2}{2} - 6x^3 + \frac{11x^4}{8} + \ldots$ 17) $1 - 18x^2 - 24x^3 + 28x^4 + \ldots$

18) $x + 2x^2 + \frac{5}{6}x^3 - x^4 + \ldots$ 19) $3x^2 + 4x^3 + \ldots$

20) $1 - \frac{81}{2}x^2 - 36x^3 + \frac{2015}{8}x^4 + \ldots + i(9x + 4x^2 - \frac{249}{2}x^3 - 162x^4 + \ldots)$

Chapter 14

FUNCTIONS OF BIPOLYNOMIALS

In this chapter we are concerned with expressing various functions of binary polynomials (bipolynomials) as power series jointly in x and y.

Just as the facility for expressing functions of polynomials as power series opens up ways of handling functions of a single variable, a similar facility opens up possibilities with functions of two variables. Simple applications of one or two of the procedures developed in this chapter are shown in Chapter 16, on the solution of partial differential equations.

The functions considered in this chapter are:

$$P(x,y) \times Q(x,y), \quad P^2(x,y), \quad \frac{1}{P(x,y)}, \quad P^{\frac{1}{2}}(x,y), \quad P(x,y) \div Q(x,y), \quad \ln P(x,y), \quad \exp P(x,y), \quad P^m(x,y),$$

$$\cos P(x,y) \ \& \ \sin P(x,y),$$

where $P(x,y)$ & $Q(x,y)$ are bipolynomials in x and y, and m is constant.

This is not essentially a difficult topic. The main problem with it is the sheer number of terms, owing to the entry of an extra dimension. As in Chapter 13 the coefficients are evaluated one by one. For brevity the whole process is referred to as **evaluating the function**.

We begin with the first of the above list, the products of bipolynomials.

PRODUCTS

Example 1   Let P = 2   + 3x   &   Q = 3   + x
                        +y   + 2xy   +2y + xy

Then we find their product is given by:

$$PQ = \quad 6 \qquad + \quad 11x \quad + \quad 3x^2$$
$$+7y \quad + \quad 15xy \quad + \quad 5x^2y$$
$$+2y^2 \quad + \quad 5xy^2 \quad + \quad 2x^2y^2$$

These coefficients are obtained from a procedure coming under the 'vertically and crosswise' sutra. The scheme for calculating the coefficients can be represented as follows, where the dots represent coefficients of P & Q:

TABLE I. Scheme for calculating coefficients of P & Q

In Table I figures at the end of each line are taken from different matrices, e.g.

for the coefficient of xy we have the pattern $\times$

Now $\diagdown$ represents $2 \times 1 + 3 \times 2$, coming from:

P = $\begin{matrix} 2 & \cdot \\ \cdot & \cdot \end{matrix}$ and Q = $\begin{matrix} \cdot & \cdot \\ & 1 \end{matrix}$, giving 2×1,

& P = $\begin{matrix} \cdot & \cdot \\ \cdot & 2 \end{matrix}$ and Q = $\begin{matrix} 3 & \cdot \\ \cdot & \cdot \end{matrix}$ which gives 2×3

Thus each line represents two products in general. Where the elements to be multiplied are in like positions, this is represented by a square in place of a dot, and yields one term, not two (cf the 'dwandwa-yoga' term when squaring).

Hence ✗ represents $2 \times 1 + 3 \times 2$
$1 \times 1 + 3 \times 2 = 15$,

which is the coefficient of xy, and ⬛ represents 3×1, the coefficient of x^2.

$$\text{If } P = \begin{matrix} a_{11} & +a_{12}x & +a_{13}x^2 \\ +a_{21}y & +a_{22}xy & +a_{23}x^2y \\ +a_{31}y^2 & +a_{32}xy^2 & +a_{33}x^2y^2 \end{matrix} \quad \text{and } Q = \begin{matrix} b_{11} & +b_{12}x & +b_{13}x^2 \\ +b_{21}y & +b_{22}xy & +b_{23}x^2y \\ +b_{31}y^2 & +b_{32}xy^2 & +b_{33}x^2y^3 \end{matrix}$$

then for $P \times Q$ we employ the following scheme:

TABLE II. Scheme for calculating coefficients of $P \times Q$.

E.g. for the coefficient of x^2 we have: $(a_{11}b_{13} + b_{11}a_{13}) + a_{12}b_{12}$,

coming from components ——— and ⬛

This application tells us something about the sutra 'vertically and crosswise'. In some applications 'vertically' can equally apply to horizontal terms, especially where the choice of which is a vertical and which a horizontal is arbitrary. Thus it appears reasonable to suppose that 'vertically' can, in such cases, be an abbreviated reference to either vertical or horizontal relationships (Sanskrit grammarians make use of a similar device, known as 'pratyahara'). Alternatively, is it stretching a point too much to suggest that 'crosswise' could be interpreted by a Latin cross, as well by a Greek cross?

APPLICATIONS

1. The product $(2.1 + 3.2x)(3.2 + 1.1x)$ can be evaluated by putting $y = 0.1$, so that $2 + y$ represents 2.1 etc., and proceeding as in Example 1 above. Then by replacing y by 0.1 at the end of the calculation, the solution emerges:

$$6.72 + 12.55x + 3.52x^2$$

This is of course, a trivial example, but it has considerable implications.

2. Likewise, the product of complex numbers can be obtained by this method, by taking $y = i = \sqrt{-1}$, and working with the first two rows only.

Hence, $[(2 + i) + (3 + 2i)x][(3 + 2i) + (1 + i)x] = 4 + 6x + 1x^2 + i(7 + 15x + 5x^2)$, since $y^2 = i^2 = -1$, so that the coefficients of the third row are subtracted from those of the first.

SQUARES

Example 2   This is a special case of a product: $A = 1 + x$
$$+y + xy$$

$$\square \ A^2 = \begin{matrix} 1 & +2x & +x^2 \\ +2y & +4xy & +2x^2y \\ +y^2 & +2xy^2 & +x^2y^2 \end{matrix}$$

The duplex (or dwandwa-yoga) process can be seen at work here.

E.g. with the central term of A^2, obtained from the cross-products we double up, to obtain $(2 \times 1 \times 1 + 2 \times 1 \times 1)xy = 4xy$.

Example 3

$$B = \begin{array}{lll} 2 & +3x & +x^2 \\ +4y & +2xy \\ +0y^2 & +3xy^2 \end{array}$$

$$B^2 = \begin{array}{|ccccc} 4 & 12 & 13 & 6 & 1 \\ 16 & 32 & 20 & 4 & 0 \\ 16 & 28 & 22 & 6 & 0 \\ 0 & 24 & 12 & 0 & 0 \\ 0 & 0 & 9 & 0 & 0 \end{array}$$

The various powers of x and y have been omitted here, but are to be understood.
The surrounding line (three sides of a square) is intended to remind us of this, and also to distinguish this as a special case of a matrix to which, as the present section is concerned with showing, various procedures apply which are not applicable to matrices in general.
It is perhaps of interest to note that in ordinary matrix notation,

$$B = \begin{pmatrix} 1 & y & y^2 \end{pmatrix} \begin{pmatrix} 2 & 3 & 1 \\ 4 & 2 & 0 \\ 0 & 3 & 0 \end{pmatrix} \begin{pmatrix} 1 \\ x \\ x^2 \end{pmatrix}$$

Sometimes we may wish to write the letter representing the array above it, in which case the shape ⌒⎯ is useful in pointing to the letter in question.

E.g.
$$\overset{R}{\begin{array}{|cc} 1 & 3 \\ 2 & 4 \end{array}} \quad \text{means} \quad R = \begin{vmatrix} 1 & 3 \\ 2 & 4 \end{vmatrix} = \begin{array}{ll} 1 & +3x \\ +2y & +4xy \end{array}$$

SQUARE ROOTS

The square root can be obtained by reversing the above procedure and working argumentally.
e.g. taking Example 2 in reverse, we have:

$$A^2 = \begin{array}{|ccc} 1 & 2 & 1 \\ 2 & 4 & 2 \\ 1 & 2 & 1 \end{array}$$

Starting in the top left-hand corner, with the coefficient of unity, we have:

$1^{1/2} = +1$ or -1

Taking the positive root initially gives:

$$+\sqrt{A^2} = \begin{pmatrix} +1 & \cdot & \cdot & \cdots \\ \cdot & \cdot & \vdots & \cdots \\ \cdots\cdots\cdots\cdots\cdots \end{pmatrix} = \begin{pmatrix} 1 & a_{12} & a_{13} \\ a_{21} & a_{22} & a_{23} \\ \cdots\cdots\cdots\cdots \end{pmatrix}$$

Working along the first row (or column), the procedure is the same as for finding the square root of a polynomial—which will be done here argumentally.

For a_{12}, we have: $2 \times 1 \times a_{12} = 2$
$$\therefore \quad a_{12} = 1, \text{ giving:}$$

$$+\sqrt{A^2} = \begin{pmatrix} 1 & 1 & a_{13} & \cdots \\ a_{21} & a_{22} & a_{23} & \cdots \\ \cdots\cdots\cdots\cdots\cdots \end{pmatrix}$$

For a_{13}, we have: $2 \times 1 \times a_{13} + 1^2 = 1$,
$$\therefore \quad a_{13} = 0$$

Similarly $\quad a_{14} = a_{15} = a_{16} = \cdots = 0$

In the second row, for a_{21} we have: $2 \times 1 \times a_{21} = 2$
$$\therefore \quad a_{21} = 1, \text{ giving:}$$

$$+\sqrt{A^2} = \begin{pmatrix} 1 & 1 & 0 & 0 & 0 & \cdots \\ 1 & a_{22} & a_{23} & & \cdots \\ \cdots\cdots\cdots\cdots\cdots \end{pmatrix}$$

For a_{22}, $2 \times 1 \times 1 + 2 \times 1 \times 1 a_{22} = 4$
$$\therefore \quad a_{22} = 1$$

Continuing, we now find:

$$a_{23} = a_{24} = \ldots = 0$$
$$\&\ a_{31} = a_{32} = \ldots = 0$$
$$\&\ a_{41} = a_{42} = \ldots = 0$$

I.e. $+\sqrt{A^2} = \begin{vmatrix} 1 & 1 \\ 1 & 1 \end{vmatrix} = \begin{matrix} 1 & +x \\ +y & +xy \end{matrix}$

Similarly, $-\sqrt{A^2} = \begin{vmatrix} -1 & -1 \\ -1 & -1 \end{vmatrix} = \begin{matrix} -1 & -x \\ -y & -xy \end{matrix}$ $\begin{vmatrix} 1 & 1 & 0 & 0 & 0 \ldots \\ 1 & a_{22} & a_{23} & & \ldots \\ & & \ldots & & \end{vmatrix}$

This argumental process can be formalised. Proceeding systematically from left to right along the rows (or columns), use the solution obtained so far to subtract the two-dimensional dwandwa-yoga from the appropriate element in the array being square-rooted, then divide by twice the top left-hand element, i.e. by a_{11}, the coefficient of unity.

Example 4   Find $B^{½}$, where $B = \begin{matrix} 4 & +12x & +9x^2 \\ +4y & +22xy & +24x^2y \\ +1y^2 & +8xy^2 & +16x^2y^2 \end{matrix}$

To take the square root, since $+\sqrt{4} = +2$,   we have, initially:

$$B^{½} = \begin{vmatrix} +2 & b_{12} & b_{13} & \ldots \\ b_{21} & b_{22} & b_{23} & \ldots \end{vmatrix}$$

Using twice the coefficient of unity i.e. 2×2 as divisor we can begin to write down the solution as follows:

$$
\begin{array}{c|ccc}
 & 4 & 12 & 9 \\
 & 4 & 22 & 24 \\
 & 1 & 8 & 16 \\
\hline
4 & 2 & &
\end{array}
$$

The initial dwandwa-yoga (or duplex) is zero, since the leading term (+2) does not contribute to it. As a reminder, it is useful to separate +2 from the rest of the solution by a diagonal stroke, as is customarily done in finding ordinary square roots by the method shown in Chapter 1.

Then $b_{12} = 12 \div 4 = 3$

Continuing the first row as for the square root of a polynomial:

$b_{13} = (9 - 3^2) \div 4 \quad = 0$
$b_{14} = (0 - 2 \times 3 \times 0) \div 4 = 0$
& $b_{15} = b_{16} = \ldots \ldots \quad = 0$

In the second row we have:

$b_{21} = 4 \div 4 \qquad\qquad = 1$
$b_{22} = (22 - 2 \times 1 \times 3) \div 4 = 4$

So far this gives:

$$
\begin{array}{r|cccccc}
 & 4 & 12 & 9 & & & \\
 & 4 & 22 & 24 & & & \\
 & 1 & 8 & 16 & & & \\
\hline
4 & 2\!\!\!\diagup & 3 & 0 & 0 & 0 & \ldots \\
 & 1 & 4 & & & &
\end{array}
$$

Continuing,
$b_{23} = (24 - 2 \times 1 \times 0 - 2 \times 3 \times 4) \div 4 \quad = 0$
$b_{24} = (0 - 2 \times 1 \times 0 - 2 \times 4 \times 0 - 2 \times 3 \times 0) \div 4 = 0$
& $b_{25} = b_{26} = \ldots . \qquad\qquad\qquad = 0$

In the third row,
$b_{31} = (1 - 1^2) \div 4 \qquad\qquad = 0$
$b_{32} = (8 - 2 \times 0 \times 3 - 2 \times 1 \times 4) \div 4 = 0$
& $b_{33} = b_{34} = \ldots . \qquad\qquad = 0$, etc.

Likewise, $b_{41} = b_{42} = \ldots . = 0$, etc.

In this way the two solutions emerge,

$$
B^{\frac{1}{2}} = \pm \left(\begin{array}{ccc} 2 & + & 3x \\ +y & + & 4xy \end{array} \right)
$$

In general the procedure leads to an infinite series of terms, jointly in powers of x and y.

Example 5   Find $C^{\frac{1}{2}}$, given: C =

$$\begin{array}{lll} 100 & +40x & +5x^2 \\ +20y & +6xy & +x^2y \\ +3y^2 & +5xy^2 \end{array}$$

Working and solution:

+20	10	2	$\frac{1}{20}$	$\frac{-1}{100}$
	1	$\frac{1}{10}$	$\frac{1}{40}$	$\frac{-9}{2000}$
		$\frac{1}{10}$	.044	.00915	. . . . . .

. . . .

i.e. $C^{\frac{1}{2}} = !$
$$\left\{ \begin{array}{llll} 10 & +2x & +\frac{1}{20}x^2 & -\frac{1}{100}x^3 + ... \\ +y & +0.1xy & +0.025x^2y & -0.0045x^3y + ... \\ +0.1y^2 & +0.044xy^2 & +0.00915x^2y^2 & +... \\ +...... \end{array} \right\}$$

In the same way the procedure for cubing bipolynomials can be used to obtain cube roots argumentally, and the process can then be formalised, as was done with square-rooting. This procedure is not shown here; instead, the more general method of obtaining powers of a polynomial is given, which method includes fractional powers.

TWO APPLICATIONS

1) Find $S = \sqrt{102.03 + 40.65x + 5.1x^2}$

Putting y = 0.1, we find that this example has been worked out in Example 5 above.
Finally, replacing y by 0.1 we have:
S = 10.101 + 2.01044x + 0.5025915x² – 0.01045x³ + ...

2) Determine $T = [(10\,\overline{3} + 20i) + (4\,\overline{5} + 6i)x + (5 + i)x^2]^{\frac{1}{2}}$ as a power series in x.

On putting y = i = $\sqrt{-1}$ we find that Example 5 above gives the solution to this problem also, and finally replacing y by i we have:

$T = !\ \{(10.\,\overline{1} + i) + (2.0\,\overline{44} + 0.1i)x + (0.05\,\overline{915} + 0.025i)x^2 - (0.05 + 0.0045i)x^3......\}$

DIVISION

$P(x,y) \div Q(x,y)$ by the 'transpose and apply' method.

The sutras used here are 'transpose and apply' and 'vertically and crosswise'.

Example 6   Divide
$$\begin{array}{llll} 2 & +4x & +0x^2 & +3x^3 \\ +3y & +xy & +2xy^2 \\ +y^2 & +3xy^2 \end{array}$$
by
$$\begin{array}{ll} 1 & +3x \\ +2y & +xy \end{array}$$

The complete working is written down as follows, with the solution appearing at bottom right:

$$
\begin{array}{c}
Q \left\{
\begin{array}{l|l}
1 + 3x & 2 + 4x + 0x^2 \\
+2y + xy & +3y + xy + 2x^2y \\
& +y^2 + 3xy
\end{array}
\right\} P \\[2em]
T \left\{
\begin{array}{cc|l}
& -3 & 2 - 2x + 6x^2 - 15x^3 \\
-2 & -1 & -1y + 6xy - 26x^2y + 102x^3y \\
& & 3y^2 - 17xy^2 - 14x^2y^2 \ldots \ldots
\end{array}
\right\} S
\end{array}
$$

This method is applicable when the constant term in the divisor is unity. There is no division to be performed, only addition and multiplication (and subtraction, when adding negative terms).

The steps are as follows, using dots to represent currently unused figures:

ROW 1:

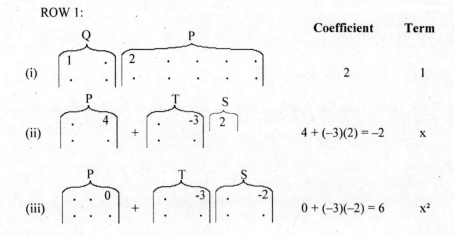

	Coefficient	Term
(i)	2	1
(ii)	$4 + (-3)(2) = -2$	x
(iii)	$0 + (-3)(-2) = 6$	x^2

Explanation of the first two steps

(i)   The top left-hand '2' in array P is written straight down as the top left-hand digit of S, the solution array. It is the coefficient of unity in the solution.

(ii)   The coefficient of x in the solution arises from adding 4 (the coefficient of x in the numerator) the product of -3 (the transposed coefficient in the first row of the operating array, T) and 2 (the first coefficient in the solution).

The steps continue:

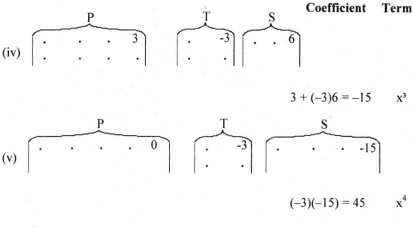

	Coefficient	Term
	$3 + (-3)6 = -15$	x^3
	$(-3)(-15) = 45$	x^4

and so on.

ROW 2:

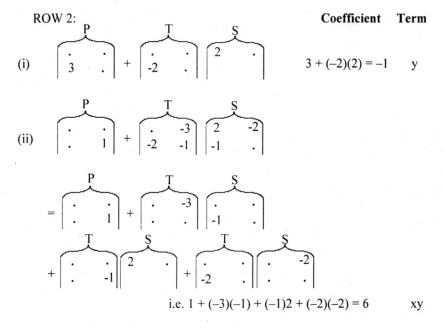

Coefficient   Term

(i) $3 + (-2)(2) = -1$ y

i.e. $1 + (-3)(-1) + (-1)2 + (-2)(-2) = 6$ xy

The foregoing can be written much more neatly by using the above array notation to place each newly-emerging solution coefficient, i.e.:

(ii)

Hereafter the explanations of the steps will be written in this form.

ROW 2 (cont.)

(iii)

$$\begin{array}{c} P \\ \begin{vmatrix} \cdot & \cdot & \cdot \\ \cdot & \cdot & 2 \end{vmatrix} \end{array} + \begin{array}{c} T \\ \begin{vmatrix} \cdot & \cdot & -3 \\ -2 & -1 \end{vmatrix} \end{array} \begin{array}{c} S \\ \begin{vmatrix} \cdot & -2 & 6 \\ \cdot & 6 \end{vmatrix} \end{array} = \begin{array}{c} S \\ \begin{vmatrix} \cdot & \cdot & \cdot \\ \cdot & \cdot & -26 \end{vmatrix} \end{array}$$

(iv)

$$\begin{array}{c} P \\ \begin{vmatrix} \cdot & \cdot & \cdot & \cdot \\ \cdot & \cdot & \cdot & 0 \end{vmatrix} \end{array} + \begin{array}{c} T \\ \begin{vmatrix} \cdot & \cdot & -3 \\ -2 & -1 \end{vmatrix} \end{array} \begin{array}{c} S \\ \begin{vmatrix} \cdot & \cdot & 6 & -15 \\ \cdot & \cdot & -26 \end{vmatrix} \end{array} = \begin{array}{c} S \\ \begin{vmatrix} \cdot & \cdot & \cdot & \cdot \\ \cdot & \cdot & \cdot & 10\vdots \end{vmatrix} \end{array}$$

Example 7   As an application, suppose we wish to evaluate:

$$\frac{2.0301 + 4.0103x + 0.02x^2 + 3x^3}{1.02 + 3.01x}$$

Putting y = 0.01 and proceeding as in Example 6 above, we have, on finally replacing y by 0.01:

$$D = 2.0\,\bar{1}\,03 - 2.0\overline{6}17x + 6.\,\overline{2614} - 14.\overline{02}x^3 + \ldots.$$

ARGUMENTAL DIVISION

The method of argumental division was explained in Chapter 1. Here, we extend its application to $P(x,y) \div Q(x,y)$.

The advantage of including the method at this point is, firstly that it explains why the previous method works, and secondly it leads naturally on to the method of 'straight division', giving an explanation of the latter en passant.

Example 8   Given

$$\begin{array}{c} P \\ \begin{vmatrix} 2 & 4 & 6 \\ 4 & 8 & 1 \end{vmatrix} \end{array} \div \begin{array}{c} Q \\ \begin{vmatrix} 2 & 1 & 1 \\ 2 & 3 \end{vmatrix} \end{array} = \begin{array}{c} S \\ \begin{vmatrix} a_0 & a_1 & a_2 \ldots \\ b_0 & b_1 & b_2 \ldots \\ c_0 & c_1 & c_2 \ldots \end{vmatrix} \end{array}$$

obtain the solution coefficients, $a_0, a_1, \ldots, b_0, b_1, \ldots$.

The procedure is, work out one coefficient at a time, proceeding in some reasonable sequence from the top left-hand corner. Let us start by working along the first row, then the second, and so on. We work out $Q \times S$ for each coefficient, and compare it with P.

The steps are as follows:

ROW 1

(i) $\left[\overbrace{2\ldots\ldots}^{P}\right] \div \left[\overbrace{2\ldots\ldots}^{Q}\right] = \left[\overbrace{1\ldots\ldots}^{S}\right]$

(ii) $\left[\overbrace{2\quad 4\ldots}^{P}\right] \div \left[\overbrace{2\quad 1\ldots}^{Q}\right] = \left[\overbrace{1\quad \frac{3}{2}\ldots}^{S}\right]$ since $2a_1 + 1 \times 1 = 4$

(iii) $\left[\overbrace{2\quad 4\quad 6}^{P}\right] \div \left[\overbrace{2\quad 1\quad 1\ldots}^{Q}\right] = \left[\overbrace{1\quad \frac{3}{2}\quad \frac{7}{4}\ldots}^{S}\right]$

since $2a_2 + 1 \times \frac{3}{2} + 1 \times 1 = 6$

(iv) $\left[\overbrace{2\quad 4\quad 6\quad 0\ldots}^{P}\right] \div \left[\overbrace{2\quad 1\quad 1\quad 0\ldots}^{Q}\right] = \left[\overbrace{1\quad \frac{3}{2}\quad \frac{7}{4}\quad \frac{-13}{8}\ldots}^{S}\right]$

since $2a_3 + 1 \times \frac{3}{2} + 1 \times \frac{7}{4} + 0 \times 1 = 0,$

And so on.

ROW 2

(i) $\left[\overbrace{\begin{matrix}\cdot \quad \cdot\\4\end{matrix}}^{P}\right] \div \left[\overbrace{\begin{matrix}2 \quad \cdot \quad \cdot\\2 \quad \cdot \quad \cdot\end{matrix}}^{Q}\right] = \left[\overbrace{\begin{matrix}1 \quad \cdot \quad \cdot\\1 \quad \cdot \quad \cdot\end{matrix}}^{S}\right]$ since $2b_0 + 2 \times 1 = 4$

(ii) $\left[\overbrace{\begin{matrix}\cdot \quad \cdot \quad \cdot\\\cdot \quad 8\end{matrix}}^{P}\right] \div \left[\overbrace{\begin{matrix}2 \quad 1\\2 \quad 3\end{matrix}}^{Q}\right] = \left[\overbrace{\begin{matrix}1 \quad \frac{3}{2}\ldots\\1 \quad \frac{1}{2}\ldots\end{matrix}}^{S}\right]$

since $2b_1 + 1 \times 1 + 3 \times 1 + 2 \times \frac{3}{2} + 3 \times 1 = 8$

(iii) $\left[\overbrace{\begin{matrix}\cdot \quad \cdot \quad \cdot\\\cdot \quad \cdot \quad 1\end{matrix}}^{P}\right] \div \left[\overbrace{\begin{matrix}2 \quad 1 \quad 1\\2 \quad 3 \quad 0\end{matrix}}^{Q}\right] = \left[\overbrace{\begin{matrix}1 \quad \frac{3}{2}\quad \frac{7}{4}\ldots\\1 \quad \frac{1}{2}\quad \frac{-17}{4}\ldots\end{matrix}}^{S}\right]$

For $\quad 2b_2 + 1 \times \frac{1}{2} + 1 \times 1$
$+ 2 \times \frac{7}{4} + 3 \times \frac{3}{2} + 0 \times 1 = 1$

And similarly for other elements of Row 2, Row 3, etc.

The coefficient to be determined is in each case multiplied by the top left-hand digit of the divisor (in this case 2). In the previous Paravartya division example, the top left-hand digit of the divisor being unity no multiplication (or subsequent division) was necessary, and by transposing signs all other terms were transferred to the right-hand side of each equation, simply leaving us with the coefficient to be determined on the left and its value on the right.

When the top left-hand digit of the divisor is not unity we use the procedure of 'straight division'. As the 'argumental division' method showed us, since the top left-hand digit of the divisor (call it 'D') is multiplied by the coefficient to be evaluated, we need to divide by D after the necessary subtractions. These latter are obtained by multiplying the flag digits by the components of the solution already obtained, as will now be shown.

STRAIGHT DIVISION

We will take the same example as was used for argumental division, differing only in that remainder terms will now be shown for powers of x above the second, and for powers of y above the first.

Example 9
$$\begin{pmatrix} 2 & +4x & +6x^2 \\ +4y & +8xy & +x^2y \end{pmatrix} \div \begin{pmatrix} 2 & +x & +x^2 \\ +2y & +3xy \end{pmatrix}$$

The complete working and solution is as follows, remainders being placed to the right of the one dashed line and below the other.

$$
\begin{array}{ccc|ccc}
\text{FLAG (F)} & & & & P & \\
- & 1 & 1 & 2 & 4 & 6 \\
2 & 3 & - & 4 & 8 & 1 \\
\hline
& & & 1 & + \tfrac{3}{2}x + \tfrac{7}{4}x^2 & - \tfrac{13}{4}x^3 - \tfrac{7}{4}x^4 \\
& & & 1y & + \tfrac{1}{2}xy - \tfrac{17}{4}x^2y & - \tfrac{3}{2}x^3y + \tfrac{17}{4}x^4y \\
& & & -2y^2 & - 4xy^2 + 7x^2y^2 & + \tfrac{51}{4}x^3y^2 + 0x^4y^2
\end{array}
$$

The flag digits consist of all digits of the divisor except the leading (upper left-hand) term.

STEPS FOR ROW 1

(i) $\left\lceil \overset{P}{\overbrace{2 \quad . \quad .}} \right\rceil \div 2 = \left\lceil \overset{S}{\overbrace{1 \quad . \quad .}} \right\rceil$

(ii) $\left\{ \left\lceil \overset{P}{\overbrace{. \quad 4 \quad .}} \right\rceil - \left\lceil \overset{F}{\overbrace{. \quad 1 \quad .}} \right\rceil \left\lceil \overset{S}{\overbrace{1 \quad . \quad .}} \right\rceil \right\} \div 2 = \left\lceil \overset{S}{\overbrace{. \quad \tfrac{3}{2} \quad .}} \right\rceil$

(iii) $\left\{ \left\lceil \overset{P}{\overbrace{. \quad . \quad 6}} \right\rceil - \left\lceil \overset{F}{\overbrace{. \quad 1 \quad 1}} \right\rceil \left\lceil \overset{S}{\overbrace{1 \quad \tfrac{3}{2} \quad .}} \right\rceil \right\} \div 2 = \left\lceil \overset{S}{\overbrace{. \quad . \quad \tfrac{7}{4}}} \right\rceil$

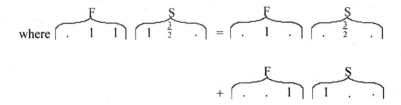

I.e. $\{6 - 1\%\frac{1}{2} - 1\%1\} + 2 = \frac{7}{4}$

For the remainders in the first row, the procedure is essentially the same as this, except that there is no division by the leading term, (2), and that reminder terms are not used to calculate further remainder terms.

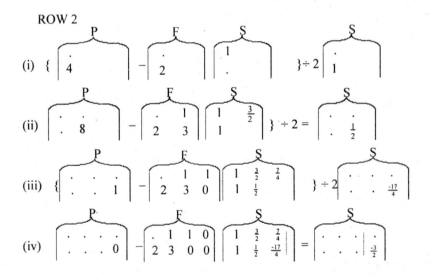

This completes the first row.

ROW 2

The features of this last step, being a calculation of a remainder-coefficient, are:

(a) no division by the divisor D (=2),

(b) remainder-digits are not multiplied by flag-digits.

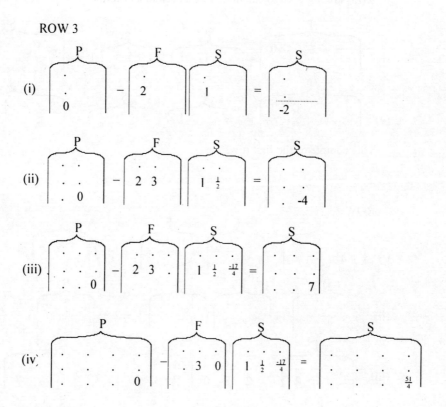

This completes the second row.

ROW 3

POWERS OF P(x,y)

The procedure for $P^m(x,y)$, where m is a constant, is a natural extension of the procedure for $P^m(x)$.

Example 10 Evaluate $\begin{pmatrix} 1 & +x \\ +2y & +xy \end{pmatrix}^3$ as a power series in x and y.

$$\text{Let } \begin{pmatrix} 1 & +x \\ +2y & +xy \end{pmatrix}^3 = \begin{pmatrix} a_0 & +b_0x & +c_0x^2 & +d_0x^3 \\ +a_1y & +b_1xy & +c_1x^2y & +d_1x^3y \\ +a_2y^2 & +b_2xy^2 & +... & \\ +a_3y^3 & +b_3xy^3 & +... & \end{pmatrix} \qquad ...(1)$$

Taking logs, differentiate Equation (1) partially w.r.t. x, and then with respect to y, and then equate coefficients, giving:

$$\frac{3\begin{pmatrix} 1 \\ +y \end{pmatrix}}{\begin{pmatrix} 1 & +x \\ +2y & +xy \end{pmatrix}} = \frac{\begin{pmatrix} \overset{3}{b_0} & \overset{6}{+2c_0x} & \overset{3}{+3d_0x^2} \\ \overset{15}{+b_1y} & \overset{24}{+2c_1xy} & \overset{9}{+3d_1x^2y} \\ \overset{24}{+b_2y^2} & \overset{30}{+2c_2xy^2} & \overset{9}{+3d_2x^2y^2} \\ \overset{12}{+b_3y^3} & \overset{12}{+2c_3xy^3} & \overset{3}{+3d_3x^2y^3} \end{pmatrix}}{\begin{pmatrix} \overset{1}{a_0} & \overset{3}{+b_0x} & \overset{3}{+c_0x^2} & \overset{1}{+d_0x^3} \\ \overset{6}{+a_1y} & \overset{15}{+b_1xy} & \overset{12}{+c_1x^2y} & \overset{3}{+d_1x^3y} \\ \overset{12}{+a_2y^2} & \overset{24}{+b_2xy^2} & \overset{15}{+c_2x^2y^2} & \overset{3}{+d_2x^3y^2} \\ \overset{8}{+a_3y^3} & \overset{12}{+b_3xy^3} & \overset{6}{+c_3x^2y^3} & \overset{1}{+d_3x^3y^3} \end{pmatrix}} \qquad ...(2)$$

And,

$$\frac{3(2+x)}{\begin{pmatrix} 1 & +x \\ +2y & +xy \end{pmatrix}} = \frac{\begin{pmatrix} \overset{6}{a_1} & +b_1x & +c_1x^2 & +d_1x^3 \\ \overset{24}{+2a_2y} & +2b_2xy & +... & \\ +3a_3y^2 & +... & & \end{pmatrix}}{\begin{pmatrix} \overset{1}{a_0} & +b_0x & +c_0x^2 & +d_0x^3 \\ \overset{6}{+a_1y} & +b_1xy & +... & \\ \overset{12}{+a_2y^2} & +... & & \\ \overset{8}{+a_3y^3} & +... & & \end{pmatrix}} \qquad ...(3)$$

The steps by which the coefficients are obtained are as follows:

Putting $x = y = 0$ in Equation (1) gives $a_0 = 1$.
Subsequent steps can be formulated in various sequences, of which one follows.

Using cross-multiplication:

Equation	Coefficient	Coefficient evaluated
(3)	unity	$a_1 = 6$
(3)	y	$2a_2 = 24$
(3)	y^2	$3a_3 = 24$
(2)	unity	$b_0 = 3$
(2)	x	$2c_0 = 6$
(2)	x^2	$3d_0 = 3$
(2)	y	$b_1 = 15$
(2)	xy	$2c_1 = 24$
(2)	x^2y	$3d_1 = 9$
(2)	y^2	$b_2 = 24$
(2)	xy^2	$2c_2 = 30$
(2)	x^2y^2	$3d_2 = 9$
(2)	y^3	$b_3 = 12$
(2)	xy^3	$2c_3 = 12$
(2)	x^2y^3	$3d_3 = 3$

$$+ \begin{pmatrix} 1 & +x \\ +2y & +xy \end{pmatrix}^3 = \begin{pmatrix} 1 & +3x & +3x^2 & +x^3 \\ +6y & +15xy & +12x^2y & +3x^3y \\ +12y^2 & +24xy^2 & +15x^2y^2 & +3x^3y^2 \\ +8y^3 & +12xy^3 & +6x^2y^3 & +x^3y^3 \end{pmatrix}$$

Example 11   Evaluate $\begin{pmatrix} 1 & +x & +2x^2 \\ & +3xy & +x^2y \end{pmatrix}^{\frac{1}{5}}$

Assume that,

$$\begin{pmatrix} 1 & +x & +2x^2 \\ & +3xy & +x^2y \end{pmatrix}^{\frac{1}{5}} = \begin{pmatrix} A_0 & +B_0x & +C_0x^2 & +D_0x^3 + ... \\ +A_1y & +B_1xy & +C_1x^2y & +... \\ +A_2y^2 & +B_2xy^2 & +... \\ +A_3y^3 & +... \\ +... \end{pmatrix} \qquad ...(4)$$

Taking logs, differentiate Equation (4) partially w.r.t. 'x', and then w.r.t. 'y', and then equate coefficients, obtaining:

$$
\dfrac{0.6\begin{bmatrix} 1 & +4x \\ +3y & +2xy \end{bmatrix}}{\begin{bmatrix} 1 & +x & +2x^2 \\ & +3xy & +x^2y \end{bmatrix}} = \dfrac{\begin{array}{l} \overset{0.6}{B_0} \quad \overset{2.16}{+2C_0x} \quad \overset{-2.568}{+3D_0x^2} \quad \overset{-0.0736}{+4E_0x^3} \; +... \\ +B_1y \quad +2C_1xy \quad +3D_1x^2y \quad +4E_1x^3y \; +... \\ +B_2y^2 \; +2C_2xy^2 \; +3D_2x^2y^2 \quad +... \\ +B_3y^3 \; +2C_3xy^3 \quad +... \\ +... \end{array}}{\begin{array}{l} \overset{1}{A_0} \quad \overset{0.6}{+B_0x} \quad \overset{1.08}{+C_0x^2} \quad \overset{-0.856}{+D_0x^3} \quad \overset{-0.0185}{+E_0x^4} \; +... \\ +A_1y \quad +B_1xy \quad +C_1x^2y \quad +D_1x^3y \quad +E_1x^4y \; +... \\ +A_4y^2 \; +B_2xy^2 \; +C_2x^2y^2 \quad +... \\ +A_3y^3 \; +B_3xy^3 \quad +... \\ +... \end{array}} \qquad ...(5)
$$

$$
\dfrac{0.6(0+3x+x^2)}{\begin{bmatrix} 1 & +x & +2x^2 \\ & +3xy & +x^2y \end{bmatrix}} = \dfrac{\begin{array}{l} \overset{0}{A_1} \quad \overset{1.8}{+B_1x} \quad \overset{-0.12}{+C_1x^2} \quad \overset{-1.176}{+D_1x^3} \; +... \\[4pt] \overset{0}{+2A_2y} \; \overset{0}{+2B_2xy} \; \overset{-2.16}{+2C_2x^2y} \; \overset{1.584}{+2D_2x^3y} \; +... \\[4pt] \overset{0}{+3A_3y^2} \; \overset{0}{+3B_3xy^2} \; \overset{0}{+3C_3x^2y^2} \; \overset{4.536}{+3D_3x^3y^2} \; +... \\[4pt] +4A_4y^3 \quad +... \\ +... \end{array}}{\begin{array}{l} \overset{1}{A_0} \quad \overset{0.6}{+B_0x} \quad \overset{1.08}{+C_0x^2} \quad \overset{0.424}{+D_0x^3} \; +... \\[4pt] \overset{0}{+A_1y} \; \overset{1.8}{+B_1xy} \; \overset{-0.12}{+C_1x^2y} \; \overset{-1.176}{+D_1x^3y} \; +... \\[4pt] \overset{0}{+A_2y^2} \; \overset{0}{+B_2xy^2} \; \overset{-1.08}{+C_2x^2y^2} \; \overset{0.792}{+D_2x^3y^2} \; +... \\[4pt] \overset{0}{+A_3y^3} \quad +... \\ +... \end{array}} \qquad ...(6)
$$

The coefficients can be evaluated as follows:

From Equation (4), putting $x = y = 0$, we have $A_0 = 1^{\frac{1}{3}}$
Take $A_0 = 1$ for the first solution.

Subsequently:

Equation	Coefficient	Coefficient evaluated
(5)	1	$B_0 = 0.6$
(5)	x	$2C_0 = 2.16$
(5)	x^2	$3D_0 = 1.272$
	.	
(6)	1	$A_1 = 0$
(6)	x	$B_1 = 1.8$
(6)	x^2	$C_1 = -0.12$
(6)	x^3	$D_1 = -1.176$
(6)	y	$2A_2 = 0$
(6)	xy	$2B_2 = 0$
(6)	x^2y	$2C_2 = -2.16$
(6)	x^3y	$2D_2 = 1.584$
(6)	y^2	$3A_3 = 0$
(6)	xy^2	$3B_3 = 0$
(6)	x^2y^2	$3C_3 = 0$

$$\square \text{ a root of} \left(\begin{array}{c} 1 \quad +x \quad +2x^2 \\ +3xy \quad +x^2y \end{array}\right)^{\frac{1}{5}} \text{ is } \left(\begin{array}{c} 1.8x \quad -0.12x^2 \quad -1.176x^3 \quad +... \\ -2.16x^2y \quad +1.584x^3y \quad +... \\ +4.536x^3y^2 \quad +... \\ +... \end{array}\right)$$

The other four roots are given by multiplying this solution by $e^{i2\pi n/5}$ where $n = 1, 2, 3$ & 4 respectively.

THE NATURAL LOGARITHM

Example 12 Obtain a power series in x and y for $\ln\left(\begin{array}{c} 1 \quad +2x \\ +3y \quad +xy \end{array}\right) = \ln\begin{vmatrix} 1 & 2 \\ 3 & 1 \end{vmatrix}$

$$\text{Assume } \ln\begin{vmatrix} 1 & 2 \\ 3 & 1 \end{vmatrix} = \begin{vmatrix} a_0 & a_1 & a_2 & a_3 & ... \\ b_0 & b_1 & b_2 & b_3 & ... \\ c_0 & c_1 & c_2 & c_3 & ... \\ ... & ... & ... & ... & ... \end{vmatrix} \quad ...(7)$$

Differentiating Equation (7) w.r.t. x:

$$\frac{\left|\begin{array}{c} 2 \\ 1 \end{array}\right|}{\left|\begin{array}{cc} 1 & 2 \\ 3 & 1 \end{array}\right|} = \begin{pmatrix} \overset{2}{a_1} & \overset{-4}{+2a_2x} & \overset{8}{+3a_3x^2} & +... \\[4pt] \overset{-5}{+b_1y} & \overset{20}{+2b_2xy} & +3b_3x^2y & +... \\[4pt] \overset{15}{+c_1y^2} & +2c_2xy^2 & +3c_3x^2y^2 & +... \\[4pt] \overset{-45}{+d_1y^3} & +... \\[4pt] +... \end{pmatrix}$$

The coefficients on the right-hand side are evaluated one at a time in the usual way. E.g. on equating coefficients of unity, $2 = 1 \times a_1$, and on equating coefficients of x, $2 \times 2 + 1 \times 2a_2 = 0$.

The coefficients b_0, c_0, etc., are evaluated by differentiating (7) w.r.t. y and equating coefficients:

$$\frac{\left|\begin{array}{cc} 3 & 1 \end{array}\right|}{\left|\begin{array}{cc} 1 & 2 \\ 3 & 1 \end{array}\right|} = \begin{pmatrix} \overset{3}{b_0} & +b_1x & +b_2x^2 & +... \\[4pt] \overset{-9}{+2c_0y} & +2c_1xy & +2c_2x^2y & +... \\[4pt] \overset{27}{+3d_0y^2} & +3d_1xy^2 & +3d_2x^2y^2 & +... \\[4pt] \overset{-81}{+4e_0y^3} & +... \end{pmatrix}$$

Re-evaluating some of the other coefficients gives a check on the working. This leaves a_0 to be calculated, and on putting $x = y = 0$ in Equation (7) we have $a_0 = 0$.

$$\square \ln\begin{pmatrix} 1 & +2x \\ +3y & +xy \end{pmatrix} = \begin{pmatrix} 0 & +2x & -2x^2 & +\tfrac{8}{3}x^3 & -4x^4 & +\tfrac{2}{5}x^5 +... \\[4pt] +3y & -5xy & +10x^2y & -20x^3y & +40x^4y & +... \\[4pt] -\tfrac{9}{2}y^2 & +15xy^2 & -\tfrac{85}{2}x^2y^2 & +110x^3y^2 & +... \\[4pt] +9y^3 & -\tfrac{135}{3}xy^3 & +\tfrac{395}{3}x^2y^3 & +... \\[4pt] -\tfrac{81}{4}y^4 & +... \\[4pt] +... \end{pmatrix}$$

Putting $y = 0.01$ in this last example, we have,

$$\ln(1.03 + 2.02x) = 0.030\ \overline{45}\ 9\ \overline{81} +2.0\ \overline{5}\ 15\ \overline{45}\ x - 2.\ \overline{1}\ 04122x^2 + 3.\ \overline{5}\ 443x^3 - 4.\ \overline{4}\ x^4 + 6.4x^5 .....$$

THE EXPONENTIAL

Example 13 Expand $\exp\left(2x + 3x^2 + y + 3xy\right)$ as a power series in x and y.

$$\text{Assume } \ln\left(\begin{array}{llll} A_0 & +A_1x & +A_2x^2 & +... \\ +B_0y & +B_1xy & +B_2x^2y & +... \\ +C_0y^2 & +C_1xy^2 & +C_2x^2y^2 & +... \\ +... \end{array}\right) = \left(\begin{array}{ll} 0 & +2x & +3x^2 \\ +y & +3xy \end{array}\right) \qquad ...(8)$$

The procedure is, we differentiate Equation (8) with respect to x, and then with respect to y. We then proceed to equate coefficients systematically—e.g. in the sequence A_0, A_1, A_2, ... , then B_0, B_1, B_2, ... , then C_0, C_1, C_2, ... , etc.

First we find A_0 from Equation (8); putting x = y = 0, we have: $A_0 = 1$

Differentiating partially with respect to x:

$$\frac{\left(\begin{array}{llll} 2 & 10 & 22 & 44\frac{2}{3} \\ A_1 & +2A_2x & +3A_3x^2 & +4A_4x^3 & +... \\ 5 & 22 & 67 & 132\frac{2}{3} \\ +B_1y & +2B_2yx & +3B_3yx^2 & +4B_4yx^3 & +... \\ 4 & 26 & 83 & 200\frac{1}{3} \\ +C_1y^2 & +2C_2y^2x & +3C_3y^2x^2 & +4C_4y^2x^3 & +... \\ +... \end{array}\right)}{\left(\begin{array}{llll} 1 & 2 & 5 & 7\frac{1}{3} \\ A_0 & +A_1x & +A_2x^2 & +A_3x^3 & +... \\ 1 & 5 & 11 & 22\frac{1}{3} \\ +B_0y & +B_1yx & +B_2yx^2 & +B_3yx^3 & +... \\ \frac{1}{2} & 4 & 13 & 27\frac{2}{3} \\ +C_0y^2 & +C_1y^2x & +C_2y^2x^2 & +C_3y^2x^3 & +... \\ +... \end{array}\right)} = \left(\begin{array}{ll} 2 & +6x \\ +3y \end{array}\right) \qquad ...(9)$$

Differentiating partially with respect to y:

$$\frac{\left.\begin{array}{l} \overset{1}{B_0} \quad \overset{5}{+B_1x} \quad \overset{11}{+B_2x^2} \quad \overset{22\frac{1}{3}}{+B_3x^3} \quad +... \\[2mm] \overset{1}{+2C_0y} \quad \overset{8}{+2C_1yx} \quad \overset{26}{+2C_2yx^2} \quad \overset{55\frac{1}{3}}{+2C_3yx^3} \quad +... \\[2mm] +3D_0y^2 \quad +3D_1y^2x \quad +3D_2y^2x^2 \quad +3D_3y^2x^3 \quad +... \end{array}\right]}{\left[\begin{array}{l} \overset{1}{A_0} \quad \overset{2}{+A_1x} \quad \overset{5}{+A_2x^2} \quad \overset{7\frac{1}{3}}{+A_3x^3} \quad +... \\[2mm] \overset{1}{+B_0y} \quad \overset{5}{+B_1yx} \quad \overset{11}{+B_2yx^2} \quad \overset{22\frac{1}{3}}{+B_3yx^3} \quad +... \\[2mm] \overset{\frac{1}{2}}{+C_0y^2} \quad \overset{4}{+C_1y^2x} \quad \overset{13}{+C_2y^2x^2} \quad \overset{27\frac{2}{3}}{+C_3y^2x^3} \quad +... \\[2mm] +... \end{array}\right]} = 1 + 3x \qquad \qquad ...(10)$$

One sequence in which the unknowns can be calculated is:

Unknown	Equation used	Coefficient of
$A_0 = 1$	(8)	unity
$A_1 = 2$	(9)	unity
$A_2 = 5$	(9)	x
$A_3 = \frac{22}{3}$	(9)	x^2
$B_0 = 1$	(10)	unity
$B_1 = 5$	(10)	x
$B_2 = 11$	(10)	x^2
$C_0 = \frac{1}{2}$	(10)	y
$C_1 = 4$	(10)	yx
$C_2 = 13$	(10)	yx^2

And so on.

Hence the solution:

$$\exp(2x + 3x^2 + y + 3xy) =$$

$$\left[\begin{array}{l} 1 \quad +2x \quad +5x^2 \quad +7\frac{1}{3}x^3 \quad +11\frac{1}{6}x^4 + ... \\[2mm] +y \quad +5yx \quad +11yx^2 \quad +22\frac{1}{3}yx^3 \quad +33\frac{1}{6}yx^4 + ... \\[2mm] +\frac{1}{2}y^2 \quad +4y^2x \quad +13y^2x^2 \quad +27\frac{2}{3}y^2x^3 \quad +50\frac{1}{12}y^2x^4 + ... \\[2mm] +... \end{array}\right]$$

COSINE AND SINE

Attention is confined here to a general method of obtaining $P(x, y)$ and $\sin P(x, y)$. A fuller investigation will doubtless yield faster methods for particular examples.

This procedure follows the same lines as that for sine and cosine of $P(x)$. We draw on De Moivre's Theorem, $\exp(i\theta) = \cos\theta + i\sin\theta$, and use the procedure for evaluating the exponential of a bivariate series. The extra consideration here is the need to evaluate the real and imaginary coefficients of each term of a bivariate series.

Example 14 Obtain $\cos(2x + 3x^2 + y + 3xy)$ and $\sin(2x + 3x^2 + y + 3xy)$ as power series expansions in x and y.

Let $\exp i(2x + 3x^2 + y + 3xy) =$

$$
\left(
\begin{array}{l}
(a_0 + iA_0) \quad +(a_1 + iA_1)x \quad +(a_2 + iA_2)x^2 \ +... \\
+(b_0 + iB_0)y \ +(b_1 + iB_1)xy \ +(b_2 + iB_2)x^2y \ +... \\
+(c_0 + iC_0)y^2 \ +(c_1 + iC_1)xy^2 \ +(c_2 + iC_2)x^2y^2 \ +... \\
+...
\end{array}
\right) \qquad ...(11)
$$

Differentiating partially with respect to x and y, and dividing these by Equation (11), we obtain two equations whose coefficients can systematically be determined by equating coefficients. The first of these is:

$$
i\left(\begin{array}{l} 2 \ +6x \\ +3y \end{array}\right) =
\frac{
\left(
\begin{array}{l}
\overset{0 \quad 2}{(a_1 + iA_1)} \quad \overset{-4 \quad 6}{+2(a_2 + iA_2)} \quad \overset{-18 \quad -4}{+3(a_3 + iA_3)} \quad +... \\[4pt]
\overset{-2 \quad 3}{+(b_1 + iB_1)y} \ \overset{-18 \quad -4}{+2(b_2 + iB_2)xy} \ \overset{-23 \quad -36}{+3(b_3 + iB_3)x^2y} \ +... \\[4pt]
\overset{-3 \quad -1}{+(c_1 + iC_1)y^2} \ \overset{-7 \quad -15}{+2(c_2 + iC_2)xy^2} \ \overset{27 \quad -52}{+3(c_3 + iC_3)x^2y^2} \ +... \\[4pt]
\overset{\frac{1}{3} \quad -\frac{1}{2}}{+(d_1 + iD_1)y^3} \qquad +... \\[4pt]
+...
\end{array}
\right)
}{
\left(
\begin{array}{l}
\overset{1 \quad 0}{(a_0 + iA_0)} \quad \overset{0 \quad 2}{+(a_1 + iA_1)x} \quad \overset{-2 \quad 3}{+(a_2 + iA_2)x^2} \quad \overset{-6 \quad -\frac{4}{3}}{+(a_3 + iA_3)x^3} \ +... \\[4pt]
\overset{0 \quad 1}{+(b_0 + iB_0)y} \ \overset{-2 \quad 3}{+(b_1 + iB_1)xy} \ \overset{-9 \quad -2}{+(b_2 + iB_2)x^2y} \ \overset{-\frac{21}{3} \quad -12}{+(b_3 + iB_3)x^3y} \ +... \\[4pt]
\overset{-\frac{1}{2} \quad 0}{+(c_0 + iC_0)y^2} \ \overset{-3 \quad -1}{+(c_1 + iC_1)xy^2} \ \overset{-\frac{7}{2} \quad -\frac{15}{2}}{+(c_2 + iC_2)x^2y^2} \ \overset{9 \quad -17\frac{1}{4}}{+(c_3 + iC_3)x^3y^2} \ +... \\[4pt]
\overset{0 \quad -\frac{1}{6}}{+(d_0 + iD_0)y^3} \ \overset{\frac{1}{3} \quad -\frac{1}{2}}{+(d_1 + iD_1)xy^3} \ +(d_2 + iD_2)x^2y^3 \qquad +... \\[4pt]
+...
\end{array}
\right)
} \qquad ...(12)
$$

The coefficients can be obtained in the following sequence (amongst others):

Coefficient	Equation	Term		Coefficient	Equation	Term
1. $a_0 = 1$	(11)	unity		7. $3a_3 = -18$	(12)	x^2
2. $A_0 = 0$	(11)	i		8. $3A_3 = -4$	(12)	ix^2
3. $a_1 = 0$	(12)	unity		9. $b_0 = 0$	(13)	unity
4. $A_1 = 2$	(12)	i		10. $B_0 = 1$	(13)	i
5. $2a_2 = -4$	(12)	x		11. $b_1 = 2$	(12)	y
6. $2A_2 = 6$	(12)	ix		12. $B_1 = 3$	(12)	iy

Differentiating partially with respect to y:

$$i(1+3x) = \cfrac{\begin{matrix} \overset{0}{}\;\overset{1}{} \\ (b_0 + iB_0) \quad +(b_1+iB_1)x \quad +(b_2+iB_2)x^2 +\ldots \\ \overset{-1}{}\;\overset{0}{} \\ +(2c_0+i2C_0)y \;+(2c_1+i2C_1)xy \;+(2c_2+i2C_2)x^2y +\ldots \\ \overset{0}{}\;\overset{-\frac{1}{2}}{} \\ +(3d_0+i3D_0)y^2 \qquad +\ldots \\ +(4e_0+i4E_0)y^3 \qquad +\ldots \\ +\ldots \end{matrix}}{\begin{matrix} \overset{1}{}\;\overset{0}{} \\ (a_0+iA_0) \quad +(a_1+iA_1)x \quad +(a_2+iA_2)x^2 +\ldots \\ \overset{0}{}\;\overset{1}{} \\ +(b_0+iB_0)y \;+(b_1+iB_1)xy \;+(b_2+iB_2)x^2y +\ldots \\ \overset{-\frac{1}{2}}{}\;\overset{0}{} \\ +(c_0+iC_0)y^2 \;+(c_1+iC_1)xy^2 \qquad +\ldots \\ \overset{0}{}\;\overset{-\frac{1}{6}}{} \\ +(d_0+iD_0)y^3 \qquad +\ldots \\ +\ldots \end{matrix}} \qquad \ldots(13)$$

Coefficient	Equation	Term		Coefficient	Equation	Term
13. $2b_2 = -18$	(12)	xy		20. $C_1 = -1$	(12)	iy^2
14. $2B_2 = -4$	(12)	ixy		21. $2c_2 = -\frac{7}{2}$	(12)	xy^2
15. $3b_3 = -23$	(12)	x^2y		22. $2C_2 = -15$	(12)	ixy^2
16. $3B_3 = -36$	(12)	ix^2y		23. $3c_3 = -27$	(12)	x^2y^2
17. $2c_0 = -1$	(13)	y		24. $3C_3 = -52$	(12)	ix^2y^2
18. $2C_0 = 0$	(13)	iy		25. $3d_0 = 0$	(13)	y^2
19. $c_1 = -3$	(12)	y^2		26. $3D_0 = -\frac{1}{2}$	(13)	iy^2

$\therefore$ we have, as solutions:

$$\cos(2x + 3x^2 + y + 3xy) = \begin{pmatrix} 1 & +0x & -2x^2 & -6x^3 & +... \\ +0y & -2xy & -9x^2y & -\frac{23}{3}x^3y & +... \\ -\frac{1}{2}y^2 & -3xy^2 & -\frac{7}{2}x^2y^2 & +9x^3y^2 & +... \\ +0y^3 & +\frac{1}{3}xy^3 & +... & \\ +... & & & \end{pmatrix}$$

and $\sin(2x + 3x^2 + y + 3xy) = \begin{pmatrix} 0 & +2x & +3x^2 & -\frac{4}{3}x^3 & +... \\ +y & +3xy & -2x^2y & -12x^3y & +... \\ +0y^2 & -xy^2 & -\frac{15}{2}x^2y^2 & -\frac{17}{3}x^3y^2 & +... \\ -\frac{1}{6}y^3 & -\frac{3}{2}xy^3 & +... & \\ +... & & & \end{pmatrix}$

CHECK:

$\sin(2x + 3x^2 + y + 3xy) + \cos^2(2x + 3x^2 + y + 3xy)$

$$= \begin{pmatrix} 1 & +0x & +0x^2 & +0x^3 & +... \\ +0y & +0xy & +0x^2y & +0x^3y & +... \\ +0y^2 & +0xy^2 & +0x^2y^2 & +0x^3y^2 & +... \\ +0y^3 & +0xy^3 & +... & \\ +... & & & \end{pmatrix}$$

= 1, verifying the correctness of the solution.

Exercise

Evaluate the following as power series of x and y:

1) $\begin{pmatrix} 3 & +2x \\ +3y & +4xy \end{pmatrix} \ominus \begin{pmatrix} 2 & +x \\ +y & +5xy \end{pmatrix}$

2) $\begin{pmatrix} 4 & +x \\ +2y & +6xy \end{pmatrix} \ominus \begin{pmatrix} 3 & +4x \\ +9y & +2xy \end{pmatrix}$

3) $\begin{pmatrix} 3 & +3x \\ +4y & +9xy \end{pmatrix}^2$

4) $\begin{pmatrix} 2 & +4x & +3x^2 \\ +4y & +2xy & +4x^2y \\ +y^2 & +3xy^2 & +2x^2y^2 \end{pmatrix} \times \begin{pmatrix} 3 & +2x & +x^2 \\ +2y & +3xy & +3x^2y \\ +y^2 & +4xy^2 & +2x^2y^2 \end{pmatrix}$

5) $\left(\begin{array}{lll} 3 & +2x & +4x^2 \\ +4y & +3xy & +3x^2y \\ +y & +2xy^2 & +3x^2y^2 \end{array} \right)^2$

6) $\left(\begin{array}{lll} 16 & +16x & +4x^2 \\ \cdot+24y & +20xy & +4x^2y \\ +9y^2 & +6xy^2 & +x^2y^2 \end{array} \right)^{\frac{1}{2}}$

7) $\left(\begin{array}{lll} 9 & +24x & +16x^2 \\ +12y & +4xy & -16x^2y \\ +4y^2 & +8xy^2 & +4x^2y^2 \end{array} \right)^{\frac{1}{2}}$

8) $\left(\begin{array}{lll} 16 & & \\ -24y & +96xy & \\ +9y^2 & -72xy^2 & +144x^2y^2 \end{array} \right)^{\frac{1}{2}}$

9) $\left(\begin{array}{lll} 9 & +12x & +4x^2 \\ -24y & -52xy & -24x^2y \\ +16y^2 & +48xy^2 & +36x^2y^2 \end{array} \right)^{\frac{1}{2}}$

10) $\left(\begin{array}{lll} 9 & -12x & +4x^2 \\ +24y & +26xy & -28x^2y \\ +16y^2 & +56xy^2 & +49x^2y^2 \end{array} \right)^{\frac{1}{2}}$

11) $\left(\begin{array}{lll} 8 & +18x & +9x^2 \\ +8y & +19xy & +12x^2y \\ +2y^2 & +5xy^2 & +3x^2y^2 \end{array} \right) + \left(\begin{array}{ll} 4 & +3x \\ +2y & +3xy \end{array} \right)$

12) $\left(\begin{array}{lll} 3 & -4x & -4x^2 \\ -5y & -xy & +6x^2y \\ -12y^2 & +13xy^2 & +4x^2y^2 \end{array} \right) + \left(\begin{array}{ll} 1 & -2x \\ = 3y & +4xy \end{array} \right)$

13) $\left(\begin{array}{llll} 3 & +2x & -7x^2 & +2x^3 \\ +7y & -13xy & +19x^2y & -x^3y \\ +3y^2 & & -8x^2y^2 & +x^3y^2 \\ +2y^3 & -9xy^3 & +13x^2y^3 & -6x^3y^3 \end{array} \right) + \left(\begin{array}{ll} 1 & +2x \\ +2y & -3xy \end{array} \right)$

14) $\left(\begin{array}{llll} 8 & +22x & +12x^2 & \\ +10y & -7xy & +4x^2y & +6x^3y \\ +y^2 & +5xy^2 & -16x^2y^2 & +4x^3y^2 \\ -y^3 & +4xy^3 & -4x^2y^3 & \end{array} \right) + \left(\begin{array}{ll} 4 & +3x \\ -y & +2xy \end{array} \right)$

15) $\begin{pmatrix} 3 & +4x & +x^2 & +2x^3 & +2x^4 \\ +4y & +17xy & +12x^2y & +2x^3y & +10x^4y \\ +4y^2 & +7xy^2 & +20x^2y^2 & +9x^3y^2 & +5x^4y^2 \\ +4y^3 & -5xy^3 & +x^2y^3 & +3x^3y^3 \\ +y^4 & -xy^4 & -3x^2y^4 & -4x^3y^4 & -2x^4y^4 \end{pmatrix} + \begin{pmatrix} 3 & -2x & +2x^2 \\ +4y & +0xy & +2x^2y \\ +y^2 & +xy^2 & +x^2y^2 \end{pmatrix}$

16) $\ln\begin{pmatrix} 1 & +3x \\ +4y & +xy \end{pmatrix}$

17) $\exp\begin{pmatrix} 4x & +3x^2 \\ +y & +4xy \end{pmatrix}$

18) $\cos\begin{pmatrix} 3x & +2x^2 \\ +3y & +4xy \end{pmatrix}$

19) $\sin\begin{pmatrix} x & -3x^2 \\ +2y & +4xy \end{pmatrix}$

20) $\exp i\begin{pmatrix} 4x & +x^2 \\ +y & +xy \end{pmatrix}$

Answers

1) $\begin{pmatrix} 6 & +7x & +2x^2 \\ +9y & +28xy & +14x^2y \\ +3y^2 & +19xy^2 & +20x^2y^2 \end{pmatrix}$

2) $\begin{pmatrix} 12 & +19x & +4x^2 \\ +42y & +43xy & +26x^2y \\ +18y^2 & +58xy^2 & +26x^2y^2 \end{pmatrix}$

3) $\begin{pmatrix} 9 & +18x & +9x^2 \\ +24y & +78xy & +54x^2y \\ +16y^2 & +72xy^2 & +81x^2y^2 \end{pmatrix}$

4) $\begin{pmatrix} 6 & +16x & +19x^2 & +10x^3 & +3x^4 \\ +16y & +28xy & +44x^2y & +31x^3y & +13x^4y \\ +13y^2 & +39xy^2 & +62x^2y^2 & +45x^3y^2 & +20x^4y^2 \\ +6y^3 & +27xy^3 & +36x^2y^3 & +35x^3y^3 & +14x^4y^3 \\ +y^4 & +7xy^4 & +16x^2y^4 & +14x^3y^4 & +4x^4y^4 \end{pmatrix}$

5) $\begin{pmatrix} 9 & +12x & +28x^2 & +16x^3 & +16x^4 \\ +24y & +34xy & +62x^2y & +36x^3y & +24x^4y \\ +22y^2 & +40xy^2 & +67x^2y^2 & +46x^3y^2 & +33x^4y^2 \\ +8y^3 & +22xy^3 & +42x^2y^3 & +40x^3y^3 & +18x^4y^3 \\ +y^4 & +4xy^4 & +10x^2y^4 & +12x^3y^4 & +9x^4y^4 \end{pmatrix}$

6) $\begin{pmatrix} 4 & +2x \\ +3y & +xy \end{pmatrix}$

7) $\begin{pmatrix} 3 & +4x \\ +2y & -2xy \end{pmatrix}$

8) $\begin{pmatrix} 4 & \\ -3y & +12xy \end{pmatrix}$

9) $\begin{pmatrix} 3 & +2x \\ -4y & -6xy \end{pmatrix}$

10) $\begin{pmatrix} 3 & -2x \\ +4y & +7xy \end{pmatrix}$

11) $\begin{pmatrix} 2 & +3x \\ +y & xy \end{pmatrix}$

12) $\begin{pmatrix} 3 & +2x \\ +4y & +xy \end{pmatrix}$

13) $\begin{pmatrix} 3 & -4x & +x^2 \\ +y & +2xy & +x^2y \\ +y^2 & -3xy^2 & +2x^2y^2 \end{pmatrix}$

14) $\begin{pmatrix} 2 & +4x & \\ +3y & -4xy & +2x^2y \\ +y^2 & -2xy^2 & \end{pmatrix}$

15) $\begin{pmatrix} 1 & +2x & +x^2 \\ & +3xy & +4x^2y \\ +y^2 & -2xy^2 & -2x^2y^2 \end{pmatrix}$

16) $\begin{pmatrix} 0 & +3x & -\frac{9}{2}x^2 & +9x^3 & -\frac{81}{4}x^4 & +\frac{243}{5}x^5 + \ldots \\ +4y & -11xy & +33x^2y & -99x^3y & +297x^4y + \ldots & \\ -8y^2 & +44xy^2 & -\frac{385}{2}x^2y^2 & +759x^3y^2 + \ldots & & \\ +\frac{64}{3}y^3 & -176xy^3 & +1012x^3y^3 + \ldots & & & \\ -64y^4 & +704xy^4 + \ldots & & & & \\ +\ldots & & & & & \end{pmatrix}$

17) $\begin{pmatrix} 1 & +4x & +11x^2 & +\frac{68}{3}x^3 + \ldots & \\ +y & +8xy & +27x^2y & +\frac{200}{3}x^3y & +\frac{779}{6}x^4y + \ldots \\ +\frac{1}{2}y^2 & +6xy^2 & +\frac{59}{2}x^2y^2 & +\frac{263}{3}x^3y^2 & +\frac{793}{4}x^4y^2 + \ldots \\ +\ldots & & & & \end{pmatrix}$

18) $\begin{pmatrix} 1 & +0x & -\frac{9}{2}x^2 & -9x^3 + \ldots & \\ +0y & -9xy & -21x^2y & +\frac{3}{2}x^3y + \ldots & \\ -\frac{9}{2}y^2 & -12xy^2 & +\frac{49}{4}x^2y^2 & +\frac{189}{2}x^3y^2 + \ldots & \\ +0y^3 & +\frac{27}{4}xy^3 + \ldots & & & \\ +\ldots & & & & \end{pmatrix}$

19) $\begin{pmatrix} 0 & +x & -3x^2 & -\frac{1}{6}x^3 + \ldots & \\ +2y & +4xy & -x^2y & +4x^3y + \ldots & \\ +0y^2 & -2xy^2 & -2x^2y^2 & +\frac{49}{3}x^3y^2 + \ldots & \\ -\frac{4}{3}y^3 & -8xy^3 + \ldots & & & \\ +\ldots & & & & \end{pmatrix}$

20) $\begin{pmatrix} 1 & +0x & -8x^2 & -4x^3 + ... \\ +0y & -4xy & -5x^2y & +\frac{29}{3}x^3y + ... \\ -\frac{1}{2}y^2 & -xy^2 & +\frac{7}{2}x^2y^2 & +10x^3y^2 + ... \\ +0y^3 & +\frac{2}{3}xy^3 + ... \\ +... \end{pmatrix} +$

$i\begin{pmatrix} 0 & +4x & +x^2 & -\frac{32}{3}x^3 + ... \\ +y & +xy & -8x^2y & -12x^3y + ... \\ +0y^2 & -2xy^2 & -\frac{9}{2}x^2y^2 & +\frac{7}{3}x^3y^2 + ... \\ -\frac{1}{6}y^3 & -\frac{1}{2}xy^3 + ... \\ +... \end{pmatrix}$

Chapter 15

THE SOLUTION OF LINEAR AND NON-LINEAR DIFFERENTIAL, INTEGRAL AND INTEGRO-DIFFERENTIAL EQUATIONS

The method of series solution of differential equations is well known. Bringing to bear the Vedic mathematical system of sutras of Sri Bharati Krishna Tirthaji makes this a very powerful approach, as is shown in the following examples.

The procedure is applicable to a wide range of problems, in particular those where boundary conditions are given at the origin or at a single point.

We start with a simple linear example to illustrate the method. Except perhaps for the lay-out the procedure of the first example is familiar to western mathematicians.

Example 1   Solve $2y + 3y' = 18 + 8x,$ ...(1)
where $y'(0) = 4$

Here, and throughout this chapter, $y' = \frac{dy}{dx}$ and $y'(0) = \frac{dy}{dx}$ evaluated with $x = 0$. An explanation of the steps will be given shortly, but the complete working and solution appears finally as follows:

$$y = a + \overset{3}{b}x + \overset{4}{c}x^2 + \overset{0}{d}x^3 + \overset{0}{e}x^4 + \dots \qquad \dots(2)$$

$$y' = \underset{4}{b} + 2\underset{0}{c}x + 3\underset{0}{d}x^2 + 4\underset{0}{e}x^3 + 5\underset{0}{f}x^4 + \dots \qquad \dots(3)$$

I.e. the solution is $\underline{y = 3 + 4x}$

Explanation: By Maclaurin's Theorem a series expansion can be used for y, as in Equation (2). Differentiating gives Equation (3), successive coefficients

of x being arranged in successive columns. Purely for purposes of explanation we will now use the following lay-out:

Coefficients		**Column**				
		1	2	3	4	
2		$y = a$	$+ bx +$	$cx^2 +$	$dx^3 + \dots$	Row 1
3		$y' = b$	$+ 2cx +$	$3dx^2 +$	$4ex^3 + \dots$	Row 2
		18	$+ 8x$			Row 3

The left-hand side of Equation (1) translates into 2 × Row 1 + 3 × Row 2, and the coefficients '2' and '3' have here been placed in a column on the left as a reminder. To facilitate the working, the right-hand side terms of Equation (1) can be placed in a third row. Thus for each successive column we have:

$$2 \times \text{Row } 1 + 3 \times \text{Row } 2 = \text{Row } 3,$$

in accordance with Equation (1). And each successive column yields the next of the unknowns, a, b, c, d, ...
The sutra here is 'by alternate elimination and retention'.

Since $y'(0) = 4$, **b = 4**, from Row 2

Write this in place, as shown in the complete solution.

To find 'a', Column 1 (absolute terms only) gives us: $2a + 3b = 18$

Since b is known to equal 4, the mental procedure is, reduce 18 by 3× 4, giving 6, and divide by 2 to obtain **a = 3**. Now write '3' above 'a' in Row 1.

So far we have:
$$\begin{array}{l} \quad\quad 3 \quad 4 \\ y = a + bx + cx^2 + \dots \\ y' = b + 2cx + 3dx^2 + \dots \\ \quad 4 \end{array}$$

Similarly, in Column 2: $2b + 3 \times 2c = 8$
$$2c = 8 - 2 \times 4 = 0$$
$$c = 0, \text{ etc.}$$

The same procedure shows that all subsequent terms are zero, and

$$\underline{y = 3 + 4x} \text{ exactly.}$$

The next example introduces a non-linearity, namely the term y^2.

Example 2 Solve $y'' + y^2 = 1 + 2x^2 + x^4$,   ...(4)

where $y(0) = 0$ and $y'(0) = 0$

The working can be written down as follows:

Column:	I	II	III	IV	V	VI	VII	VIII	IX	X
	0	0	$\frac{1}{2}$	0	$\frac{1}{6}$	0	$\frac{1}{40}$	0	$-\frac{1}{336}$	0

$$y = a + bx + cx^2 + dx^3 + ex^4 + fx^5 + gx^6 + hx^7 + ix^8 + jx^9 + .. \quad ...(5)$$
$$y'' = 2c + 6dx + 12ex^2 + 20fx^3 + 30gx^4 + 42hx^5 + 56ix^6 + 72jx^7 + ... \quad ...(6)$$

I	II	III	IV	V	VI	VII	VIII
1	0	2	0	$\frac{3}{4}$	0	$-\frac{1}{6}$	0

Explanation: From the boundary conditions we have $a = 0$ and $b = 0$.
As before, we work with coefficients of x, in succession, each step yielding another of the values of b, c, d, ...
To conform with the left-hand side of Equation (4), the successive terms for y^2 are given by the successive duplexes (or dwandwa-yogas, see Chapter 1 for the term 'duplex') of Equation (5).

For 'c': From Column 1, $2c + a^2 = 1$
since we are eliminating all except the absolute terms. And since $a^2 = 0$, the mental procedure is, reduce 1 by 0 to obtain the value of 2c. Hence we put $2c = 1$ and $c = \frac{1}{2}$, as shown in the complete solution.

For 'd': We now eliminate all terms from Equation (1) except for the coefficients of x. The dwandwa-yoga of y^2 here is $2ab = 2 \times 0 \times 0$, and from Column II we have, $2ab + 6d = 0$, and therefore $d = 0$.

For 'e': (Column III) The dwandwa-yoga is now $2ac + b^2$, and we have, $2 \times 0 \times \frac{1}{2} + 0^2 + 12e = 2$, so we write '2' above 12e, and '$\frac{1}{6}$' above e.

For 'f': (Column IV) Since the coefficient of x^3 in Equation (1) is zero, the term in y'' equals minus the dwanda-yoga term, i.e. $-2ad - 2bc$, which is zero; hence $f = 0$. And so the solution continues. Had y(0) not been zero there would have been two solutions. As it is, the two solutions coincide.

The next example has two solutions, both of which are given. The boundary conditions are given at $x = 1$, and so a series expansion in powers of $(x - 1)$ is used.

Example 3 Solve $2y + (y')^2 = 13 + 18x + 6x^2$, where $y(1) = 6$

The boundary conditions not being given at the origin, we have a choice between expanding y as a power series in $(x - 1)$, or changing to a new independent variable, $z = x - 1$. Both approaches amount to the same thing.

Following the former course, we have:

$$2y + (y')^2 = 37 + 30(x - 1) + 6(x - 1)^2, \qquad \text{where } y(1) = 6$$

The solution proceeds:

$$
\begin{array}{cccccc}
6 & 5 & 1 & 0 & 0 \\
y = A + B(x-1) + C(x-1)^2 + D(x-1)^3 + E(x-1)^4 + F(x-1)^5 + \ldots \\
y' = B + 2C(x-1) + 3D(x-1)^2 + 4E(x-1)^3 + 5F(x-1)^4 + 6G(x-1)^5 + \ldots \\
5 & 2 & 0 & 0
\end{array}
$$

We first find $A = 6$, then $B^2 = 25$
This leads to two solutions, one with $B = +5$ (given above), and one with $B = -5$.

Enclosing the first solution by horizontal lines, the second solution is written down, outside them, thus:

$$
\begin{array}{cccccc}
\underline{6 \quad -5} & \underline{-2} & \underline{0.2} & \underline{-0.11} & & \\
6 \quad 5 & 1 & 0 & 0 & 0 \\
y = A + B(x-1) + C(x-1)^2 + D(x-1)^3 + E(x-1)^4 + F(x-1)^5 + \ldots \\
y' = B + 2C(x-1) + 3D(x-1)^2 + 4E(x-1)^3 + 5F(x-1)^4 + 6G(x-1)^5 + \ldots \\
\underline{+5 \quad 2} & \underline{0} & \underline{0} & \underline{0} & \underline{0} \\
-5 \quad -4 & 0.6 & -0.44
\end{array}
$$

The two solutions are:

$$\underline{y = 6 + 5(x - 1) + (x - 1)^2} \quad \text{(exact solution)}$$

$$\underline{\&. \ y = 6 - 5(x - 1) - 2(x - 1)^2 - 0.2(x - 1)^3 - 0.11(x - 1)^4 \ldots}$$

Example 4   Solve $2y + (y')^2 + 3y'' = 2\exp(x) + \exp(2x)$ ...(7)
where $y(0) = 2, \ y'(0) = 1$

Here the presence of $(y')^2$ might us to expect two solutions, but being given $y'(0) = 1$, there is only one solution. We first expand the right-hand side of Equation (7) as a power series in x:

$$2y + (y')^2 + 3y'' = 3 + 4x + 3x^2 + \frac{10}{3!}x^3 + \frac{18}{4!}x^4 + \frac{34}{5!}x^5 + \frac{66}{6!}x^6 + \frac{130}{7!}x^7 + \ldots$$

$$
\begin{array}{cccccccc}
& 2 & 1 & 0 & \frac{1}{9} & \frac{7}{108} & & \\
\text{Let} \quad y = & A + Bx + Cx^2 + Dx^3 + Ex^4 + Fx^5 + Gx^6 + Hx^7 + \ldots
\end{array}
$$

$$
\begin{array}{ccccccccc}
& 1 & 0 & \frac{1}{3} & \frac{7}{24} & & & & \\
y' = & B + 2Cx + 3Dx^2 + 4Ex^3 + 5Fx^4 + 6Gx^5 + 7Hx^6 + 8Ix^7 + \ldots
\end{array}
$$

$$\overset{0 \quad \frac{2}{3} \quad \frac{7}{9}}{y'' = \; 2C + 6Dx + 12Ex^2 + 20Fx^3 + 30Gx^4 + 42Hx^5 + 56Ix^6 + \ldots}$$

i.e. the solution is $\underline{y = 2 + x + 0.\,\dot{1}\,x^3 + 0.064814815x^4 + \ldots}$

Example 5 Solve $y + 2y'y'' = \exp(x)\cos2x$

where $y(0) = 1$, $y'(0) = 1$ and $y'''(0) = 0$

A power series expansion for $\exp(x)\cos x$ can be obtained from MacLaurin's Theorem. Alternatively, as shown below, write down the expansion for $\exp(x)$ and below it write the expansion for $\cos2x$, keeping like powers in the same column

$$y + 2y'y'' = \left(1 + x + \tfrac{x^2}{2!} + \tfrac{x^3}{3!} + \tfrac{x^4}{4!} + \tfrac{x^5}{5!} + \tfrac{x^6}{6!} + \ldots\right)$$
$$\left(1 + 0 - \tfrac{4x^2}{2!} + 0 + \tfrac{16x^4}{4!} + 0 - \tfrac{64x^6}{6!} + \ldots\right)$$

The right-hand side coefficients of the successive powers of x can now be worked out as needed, using the 'vertically and crosswise' procedure.

Solution

$$\overset{1 \quad\;\; 1 \quad\;\; 0 \quad\quad 0 \quad\quad -\frac{1}{8} \quad -\frac{11}{240} \quad -\frac{7}{840}}{\text{Let}\quad y \;=\; a \;+\; bx \;+\; cx^2 \;+\; dx^3 \;+\; ex^4 \;+\; fx^5 \;+\; gx^6 \;+\; \ldots}$$

$$\overset{1 \quad\;\; 0 \quad\;\; 0 \quad\quad -\frac{1}{2} \quad -\frac{11}{48} \quad -\frac{7}{140}}{y' \;=\; b \;+\; 2cx \;+\; 3dx^2 \;+\; 4ex^3 \;+\; 5fx^4 \;+\; 6gx^5 \;+\; 7hx^6 + \ldots}$$

$$\overset{0 \quad\;\; 0 \quad\;\; -\frac{3}{2} \quad -\frac{11}{12} \quad\quad -\frac{7}{28}}{y'' \;=\; 2c + 6dx + 12ex^2 + 20fx^3 + 30gx^4 + 42hx^5 + 56ix^6 + \ldots}$$

i.e. the solution is $\underline{y = 1 + x - \tfrac{1}{8}x^4 - \tfrac{11}{240}x^5 - \tfrac{7}{840}x^6 + \ldots}$

The procedure for evaluating successive coefficients does not, in this case, yield a value for d, since we have $0 \times d = 0$, which is satisfied by any finite value of d.

However, the boundary condition $y'''(0) = 0$ yields $d = 0$, so avoiding an infinite number of solutions owing to d being undetermined.

Example 6   Solve $y^{\frac{1}{2}}y' + (y')^2 = 6(2 + x)^2$ ...(8)
given $y(0) = 4$

We make use of the Vedic method for finding the square root of a polynomial or power series. This is essentially the same as that for finding the square root of a number in decimal form—see Chapter 13.

Initially we assume a power series for y, and allow the successive terms of $y^{\frac{1}{2}}$ to unfold above it, the first derivative being placed below. Since $y(0) = 4$, **a = 4** and the first term for $y^{\frac{1}{2}}$ is +2 (or –2, leading to another solution).

The solution proceeds:

$$y^{\frac{1}{2}} = \underset{4}{+2} \ldots$$

$$y = a + bx + cx^2 + dx^3 + ex^4 + fx^5 + \ldots$$

$$y' = b + 2cx + 3dx^2 + 4ex^3 + 5fx^4 + 6gx^5 + \ldots$$

Now equation (8) can be rewritten, $y'(y^{\frac{1}{2}} + y') = 24 + 24x + 6x^2$ ...(9)

Considering only absolute terms, we have,

$$b(2 + b) = 24$$

i.e. we seek numbers differing by 2, whose product is 24. Clearly 4 & 6 or –4 & –6 fit the bill (hence a doubling up of the number of solutions).
Using 4 and 6 we have **b = 4**.

The solution continues:

$$y^{\frac{1}{2}} = \underset{4 \quad 4}{2 + 1x} + \ldots \qquad \text{Row 1}$$

$$y = \underset{4 \quad 2}{a + bx} + \ldots \qquad \text{Row 2}$$

$$y' = b + 2cx + 3dx^2 + \ldots \qquad \text{Row 3}$$

We are now ready to consider the coefficient of x, in Equation (9).
The left-hand side of Equation (9) is satisfied by (mentally) adding Row 3 to Row 1, and multiplying crosswise by Row 3. i.e.

$$y^{\frac{1}{2}} + y' = 6 + (1+2c)x + \qquad \text{Row 1 + Row 3}$$

$$\times$$

$$y' = 4 + 2cx + \ldots \qquad \text{Row 3}$$

the coefficient of x is $(1+2c)×4 + 6×2c$. And this must equal 24 (from the right-hand side of Equation (9)).

i.e. $4(1+2c) + 6 × 2c = 24$

∴ **c = 1**

Thus the solution so far is

$$y^{\frac{1}{2}} = \underset{4}{2} + \underset{4}{1}x + \underset{1}{0}x^2 + \ldots$$
$$y = \underset{4}{a} + \underset{2}{b}x + cx^2 + \ldots$$
$$y' = b + 2cx + 3dx^2 + \ldots$$

For on continuing the square-rooting process, the next coefficient (that of x^2) is zero.

Seeking now the coefficients of x^2 in Equation (9) we have,

$y^{\frac{1}{2}} + y'$ gives $6 + 3x + 3dx^2$
 &   y' gives $4 + 2x + 3dx^2$

Hence, proceeding vertically and crosswise, the coefficient of x^2 comes to $6(3d) + 3×2 + 4(3d) = 6$,

where the right-hand side '6' is the coefficient of x^2 on the right in Equation (9).

Hence **d = 0**.

Continuing in this way we find that

$$y = 4 + 4x + x^2 = (2 + x)^2 \text{ exactly.}$$

This is one of four solutions.

Example 7 Solve $y + \exp\left(\dfrac{dy}{dx}\right) = 1 + 3x + 2x^2$ ...(10)

given $y'(0) = 0$

Solution
Letting $y = a + bx + cx^2 + \ldots$,
Equation (10) can be rewritten.

$$\exp(b + 2cx + 3dx^2 + 4ex^3 + \ldots) = 1 + 3x + 2x^2 - y$$
$$= 1 - a + (3 - b)x + 2 - c)x^2 - dx^3 - ex^4 - \ldots \quad ...(11)$$

Differentiating Equation (11) w.r.t x and dividing by Equation (11), we have:

$$\begin{array}{ccccc} 3 & -8 & 26\frac{1}{2} & -88\frac{1}{3} & 291\frac{1}{8} \\ \end{array}$$
$$2c + 6dx + 12ex^2 + 20fx^3 + 30gx^4 + \ldots$$

$$= \frac{\overset{3 \qquad 1 \qquad 4 \qquad -8\frac{5}{6} \qquad 22\frac{1}{12} \qquad -58\frac{9}{40}}{(3-b) + (4-2c)x - 3dx^2 - 4ex^3 - 5fx^4 - 6gx^5 - \ldots}}{\underset{1 \qquad 3 \qquad \frac{1}{2} \qquad \frac{4}{3} \qquad -2\frac{5}{24} \qquad 4\frac{5}{12} \qquad -9\frac{169}{240}}{(1-a) + (3-b)x + (2-c)x^2 - dx^3 - ex^4 - fx^5 - gx^6 - \ldots}} \qquad \ldots(12)$$

Here the values of the various coefficients have been put in position. The steps by which they are obtained follow the procedure initially outlined in Chapter 7, and which, as applied here, can be summarised as follows:

Firstly, **b = 0** from boundary conditions, and then:

Step	Coefficient considered	Equation	Unknown being determined
2	unity	(11)	**a = 0**
3	unity	(12)	$2c = 3$
4	x	(12)	$6d = 3$
5	x^2	(12)	$12e = 29\frac{1}{2}$
6	x^3	(12)	$20f = -8\frac{5}{6} - 26\frac{1}{2} \triangle 3 = -88\frac{1}{3}$
7	x^4	(12)	$30g = 22\frac{1}{2} + 6\frac{5}{8} + 10\frac{2}{3} - 13\frac{1}{4} + 265 = 291\frac{1}{8}$

The solution therefore runs:

$$y = \frac{3}{2}x^2 - \frac{4}{3}x^3 + 2\frac{5}{24}x^4 - 4\frac{5}{12}x^5 + 9\frac{169}{240}x^6 + \ldots$$

Example 8   Find a real solution of: $y + \left(\frac{dy}{dx}\right)^{4.2} = 3 + x^2$ $\ldots(13)$

given $y'(0) = 1$

$$\overset{2 \qquad 1 \qquad -\frac{1}{8.4} \qquad \frac{1.03}{4.2^2}}{\text{Assume} \quad y = a_0 + b_0 x + c_0 x^2 + d_0 x^3 + \ldots} \qquad \ldots(14)$$

$$\overset{1 \qquad -1 \qquad \frac{9.4}{8.4}}{\square \ (y^{\mathfrak{R}})^{4.2} = (3 - a_0) - b_0 x + (1 - c_0)x^2 - d_0 x^3 - e_0 x^4 - \ldots}$$

Take logs and differentiate:

$$-1 \quad \tfrac{6.2}{4.2}$$

$$\square \quad \frac{4.2(2c_0 + 6d_0x + 12e_0x^2 + 20f_0x^3 + 30g_0x^4 + ...)}{b_0 + 2c_0x + 3d_0x^2 + 4e_0x^3 + 5f_0x^4 + ...}$$

$$1 \quad \tfrac{1}{4.2} \quad \tfrac{3.1}{4.2^2}$$

$$-1 \quad \tfrac{9.2}{4.2} \quad -\tfrac{3.1}{4.2^2}$$

$$= \frac{-b_0 + (2-2c_0)x - 3d_0x^2 - 4e_0x^3 - 5f_0x^4 - ...}{(3-a_0) - b_0x + (1-c_0)x^2 - d_0x^3 - e_0x^4 - ...} \qquad ...(15)$$

$$1 \quad -1 \quad \tfrac{9.4}{8.4} \quad -\tfrac{1.03}{4.2^2}$$

Equation	Steps	Boundary Conditions
(14)	$b_0=1$	since $y'(0) = 1$
(13)	since $y'(0)=1$	$\left(\frac{dy}{dx}\right)^{4.2} =1$ (principal root—other solutions
	$y(0)=2$, i.e. $a_0=2$	are complex)

Equating coefficients:

Equation	Coefficient of:	Coefficients obtained
(15)	unity	$8.4c_0 = -1$, $c_0 = -\frac{1}{8.4}$
(15)	x	$4.2 \times 6d_0 = \frac{6.2}{4.2}$, $d_0 = \frac{1.03}{4.2^2}$

DIGRESSION: DIFFERENTIATION OF RATIOS OF POLYNOMIALS

The next differential equation to be considered uses the differential of a ratio of polynomials. It will be useful, therefore, to consider this procedure next.

The process in some ways resembles a 'vertically and crosswise' product. Applying the quotient rule, $d\left(\frac{u}{v}\right) = (vdu - udv) \div v^2$, term by term, we find that $\frac{du}{dx}$ leads to multiplication by one power, and $\frac{dv}{dx}$ by another.

Example (i)   Differentiate $z = \dfrac{4+5x+x^2+2x^3}{1+2x+3x^2+4x^3}$

$$\frac{dz}{dx} = \{(1\%5 - 4\%2)(1-0) + (1\%1 - 4\%3)(2-0)x + [(1\%2 - 4\%4)(3-0) +$$
$$(2\%1 - 5\%3)(2-1)]x^2 +(2.2 - 5.4)(3-1)x^3 + (3.2 - 1.4)(3-2)x^4\}$$
$$+ (1 + 2x + 3x^2 + 4x^3)^2$$

$$= \frac{-3-22x-55x^2-32x^3+2x^4}{(1+2x+3x^2+4x^3)^2}$$

Note that "vertical" terms cancel out, because the difference in powers is zero.

Example (ii) Differentiate $z = \dfrac{3+2x+4x^2}{2+5x+3x^2}$

$$\frac{dz}{dx} = \frac{(2.2-3.5)(1-0)+(2.4-3.3)(2-0)x+(5.4-2.3)(2-1)x^2}{(2+5x+3x^2)^2} = \frac{-11-2x+14x^2}{(2+5x+3x^2)^2}$$

A similar and inevitably simpler procedure applies to the differentiation of products. The reader might be interested to note, at this point, that Tirthaji tells us that the vedic mathematicians did not advocate or even countenance any fixed sequence for expanding the subject. The implication is that a digression such as this is perfectly in the spirit of the system.

Example 9 Solve $\exp(2y) + \exp(y') = 2 + x^2$ ...(16)

given $y(0) = 0$ (whence $y'(0) = 0$, from Equation (16))

Let $\exp(2y) = a + bx + cx^2 + dx^3 + ex^4 + fx^5 + \ldots$ ...(17)

Substitute in Equation (16):

$\exp(y') = (2-a) - bx + (1-c)x^2 - dx^3 - ex^4 - \ldots$ ...(18)

Differentiate Equation (17) and divide by Equation (17):

$$2y' = \frac{\overset{0}{b} + 2cx + 3dx^2 + 4ex^3 + 5fx^4 + \ldots}{\underset{1 \qquad\; 0}{a + bx + cx^2 + dx^3 + ex^4 + \ldots}}$$ ...(19)

Substitute Expression (19) for y' in Equation (18), differentiate, and divide by Equation (18) to obtain Equation (20).

Note that on operation has been deferred. Since we can differentiate a quotient term by term, it will pay us to do so at the same time as equating coefficients.

Hence the numerator of the left-hand side of Equation (20) is initially left blank, and we have:

$$\frac{[\quad .\qquad .\qquad .\qquad]}{2(a+bx+cx^2+dx^3+ex^4+\ldots)^2} = \frac{-b+(2-2c)x-3dx^2-4ex^3-5fx^4-\ldots}{(2-a)-bx+(1-c)x^2-dx^3-ex^4-\ldots}$$ ...(20)

Since $y(0) = 0$, Equation (17) gives **a = 1**

Since $y'(0) = 0$, Equation (19) gives $\mathbf{b = 0}$

We now work with Equations (19) and (20), obtaining the differentials of the former, term by term, for insertion in Equation (20), on the top left-hand side.

Equating coefficients in Equation (20) we obtain successively:

(i) the values of those coefficients and
(ii) the terms in the numerator of the left-hand side (l.h.s.) of Equation (20).

The steps can be summarised as follows:

Term	Next coefficient in numerator of l.h.s. of Equation (20)	Unknown obtained
Unity	$1 \times 2c - 0 \times 0 = 2c$	$2c = 0$
x	$(1 \times 3d - 0 \times 0)(2 - 0) = 6d$	$6d = 4$

So far we will have written down, for Equation (20):

$$\frac{\overset{0\quad 4}{2c+6dx}}{\underset{1\ 0\ \ 0}{2(a+bx+cx^2+dx^3+ex^4+...)^2}} = \frac{\overset{0\quad\ \ 2}{-b+(2-2c)x-3dx^2-4ex^3-5fx^4-...}}{\underset{1\qquad 0}{(2-a)-bx+(1-c)x^2-dx^3-ex^4}} \qquad ...(20)$$

And in Equation (19), we will have reached the following point:

$$2y' = \frac{\overset{0\quad\ \ 0\qquad 2}{b + 2cx + 3dx^2 + 4ex^3 + 5fx^4 + ...}}{\underset{1\quad\ 0\qquad 0\qquad \frac{2}{3}}{a + bx + cx^2 + dx^3 + ex^4 + ...}}$$

Checking the last step by cross-multiplying to obtain the coefficient of x, bearing in mind that the l.h.s. denominator is squared, we have:

$0 \times 0 + 4 \times 1 = 2[1 \times 2 + 2(1 \times 0) \times 1]$, which is clearly correct.

The steps continue:

Term	Next coefficient in numerator of l.h.s. of Equation (20) (from Equation (19))	Unknown obtained from Equation (20)
x^2	$(\%4e - 0\%\frac{2}{3})(3 - 0) + (0\%2 - 0\%0)(2 - 1) = 12e$	$1 \times 12e + 0 \times 4 + 1 \times 0 =$ $(-2)1^2 + 2 \times 2 \times 2 \times 1 \times 0$ $+ 0 \times 2 \times 2 \times 1 \times 0 + 0 \times 2 \times 0^2$ $\therefore\ 12e = -4$

$$x^3 \qquad\qquad 5f(4-0) + 0(3-1) = 20f \qquad\qquad\qquad 20f+4 = 2 \times \tfrac{4}{3}$$

$$\square\; 20f = -\tfrac{4}{3}$$

The information recorded against Equations (19) and (20) is now as follows:

$$
\begin{array}{cccccc}
0 & \;\;0 & \;\;2 & -\tfrac{4}{3} & -\tfrac{1}{3} & \\
b & +\;2cx & +\;3dx^2 & +\;4ex^3 & +\;5fx^4 & +\;6gx^5+... \\
\hline
2y' = a & +\;bx & +\;cx^2 & +\;dx^3 & +\;ex^4 & +\;fx^5+... \\
1 & \;\;0 & \;\;0 & \tfrac{2}{3} & -\tfrac{1}{3} & -\tfrac{2}{30}
\end{array}
$$

$$
\begin{array}{l}
\quad 0\;\;4 \quad -4 \quad -\tfrac{4}{3} \\
\dfrac{2c+6dx+12ex^2+20fx^3+...}{2(a+bx+\;cx^2\;+\;dx^3\;+\;ex^4+...)^2} \\
\quad 1\;\;0 \qquad 0 \qquad \tfrac{2}{3} \qquad -\tfrac{1}{3}
\end{array}
\;=\;
\begin{array}{l}
\quad 0 \quad\;\; 2 \quad\;\; -2 \quad \tfrac{4}{3} \\
\dfrac{-b+(2-2c)x-3dx^2-4ex^3-5fx^4-...}{(2-a)-bx+(1-c)x^2-dx^3-ex^4-fx^5-...} \\
\quad 1 \quad\; 0 \quad\;\; 1 \quad -\tfrac{2}{3} \quad \tfrac{1}{3} \quad \tfrac{2}{30}
\end{array}
$$

The solution, as obtained so far, is obtained by taking the logarithm of Equation (17), giving:

$$y = \tfrac{1}{2}\ln\!\left(1 + \tfrac{2}{3}x^3 - \tfrac{1}{3}x^4 - \tfrac{1}{15}x^5 + ...\right)$$

INTEGRAL EQUATIONS

The same method can be applied to integral equations, as the following example illustrates.

Example 10   Solve $\displaystyle\int_0^x\!\!\int_0^x y\,dx\,dx + 2\int_0^x y\,dx = \tfrac{x^3}{6} + 2x^2 + 4x$

The working can be written down as follows:

$$
\begin{array}{ccccc}
2 & 1 & 0 & 0 & 0 \\
\end{array}
$$
Assume $y = a + bx + cx^2 + dx^3 + ex^4 + \ldots$

Integrating,

$$
\begin{array}{cccc}
2 & \tfrac{1}{2} & 0 & 0
\end{array}
$$
$$\int_0^x y\,dx \;=\; ax + \tfrac{b}{2}x^2 + \tfrac{c}{3}x^3 + \tfrac{d}{4}x^4 + ...$$

$$\int_0^x\!\!\int_0^x y\,dx \;=\; \qquad \tfrac{ax^2}{2} + \tfrac{bx^3}{6} + \tfrac{cx^4}{12} + ...$$
$$
\begin{array}{ccc}
1 & \tfrac{1}{6} & 0
\end{array}
$$

Solution: $\underline{y = 2 + x \quad \text{exactly}}$

Next consider a non-linear integro-differential equation.

Example 11 Solve $y^2 + 2\left(\frac{dy}{dx}\right)^2 + y \int_0^x ydx = 4 + 4x + 60x^2 - 8x^3 + 9x^4 + 3x^5$

where $y(0) = 2$, $y'(0) = 0$

Assume $\quad y \overset{2}{=} a + bx \overset{0}{+} cx^2 + dx^3 + ex^4 + fx^5 + gx^6 \ldots$

$\therefore \quad y' = \overset{0}{b} + 2cx + 3dx^2 + 4ex^3 + 5fx^4 + 6gx^5 + 7hx^6 + \ldots$

& $\quad \int_0^x ydx = 0 + ax + \overset{2}{\frac{b}{2}}x^2 + \overset{0}{\frac{c}{3}}x^3 + \frac{d}{4}x^4 + \frac{e}{5}x^5 + \frac{f}{6}x^6 + \ldots$

Boundary conditions lead to $a = 2$ and $b = 0$, as shown above. Succeeding steps involve the solution of simple equations as follows:

Coefficient	Equation		
x^2	$4c + 8c^2$	$= 60,$	$c = -3 \text{ (or} \frac{5}{2})$
x^3	$4d + 72d - 8$	$= -8,$	$d = 0$
x^4	$4e + 9 - 48e$	$= 9,$	$e = 0$
x^5	$4f - 60f + 3$	$= 3,$	$f = 0$
x^6	$4g - 144g$	$= 0,$	$g = 0$

Hence one solution is $\underline{y = 2 - 3x^2}$

Putting $c = \frac{5}{2}$ leads to a second solution.

TRANSCENDENTAL EQUATIONS

A special case of equations such as those presented so far is one containing no differentials, but simple terms in y, and functions of y, and functions of x. This is a type of transcendental equation, differing from those of Chapter 11 in having two variables, one being considered to be dependent upon the other.

Example 12 Solve $\quad y - \ln y = 1 - x^2$ $\quad\quad\quad\quad\quad\quad\quad\quad\quad ...(21)$

where $y = y(x)$

Assume that $\quad y = a + bx + cx^2 + dx^3 + ex^4 + \ldots$ $\qquad$ ...(22)

Substituting in Equation (22) and differentiating w.r.t. x:

$$\frac{\overset{0\ \ -1\ \ \ 0\ \ \ \frac{1}{4}\ \ \ 0\ \ \ \frac{1}{32}}{b+2cx+3dx^2+4ex^3+5fx^4+6gx^5+\ldots}}{\underset{1\ \ \ 0\ \ \ -\frac{1}{2}\ \ \ 0\ \ \ \frac{1}{16}\ \ \ 0\ \ \ \frac{1}{192}}{a+bx+cx^2+dx^3+ex^4+fx^5+gx^6+\ldots}} = \underset{0\quad -1\quad 0\quad -\frac{1}{4}\quad 0\quad -\frac{1}{32}}{-b-(2+2c)x-3dx^2-4ex^3-5fx^4-6gx^5-\ldots}\quad\ldots(23)$$

Steps		Equation used
By observation,	$y(0) = 1$ is a solution*	(21)
	$a = 1$	(22)
Equate coefficients of unity:	$b = 0$	(23)
Equate coefficients of x:	$4c = -2$	(23)
Equate coefficients of x^2:	$3d = 0$	(23)
Equate coefficients of x^3:	$8e = \frac{1}{2}$	(23)
Equate coefficients of x^4:	$10f = 0$	(23)
Equate coefficients of x^5:	$12g = \frac{1}{16}$	(23)

A solution is $\quad \underline{y = 1 - \frac{1}{2}x^2 + \frac{1}{16}x^4 + \frac{1}{192}x^6 + \ldots}$

* This is the only real solution of $y - \ln y = 1$

Exercises

Solve the following equations:

1) $\quad y' + 2y + 6 = 4x + 16$ $\qquad\qquad\qquad\qquad\qquad$ $[y = 4 + 2x]$

2) $\quad y^{\Re} + \left(\frac{dy}{dx}\right)^2 = 676x^4 + 624x^3 + 378x^2 + 108x + 20$ $\qquad$ $[y = 2 + 4x + 6x^2 + 13x^3]$
3) $\quad y'' + y' - y = -10x^4 + 38x^3 + 126x^2 + 8x + 4$ $\qquad$ $[y = 4x + 2x^3 + 10x^4]$
4) $\quad y' - 3y'' + 6 = 4 - 14x - 32x^2 - 24x^3$ $\qquad\qquad$ $[y = (1 + 2x)^3]$
5) $\quad y'' + 2y' + 6y = 14x + 4$ $\qquad\qquad\qquad\qquad\qquad$ $[y = 1 + 4x]$
6) $\quad 2y + 3y' + 6y^2 = 5\exp(x) + 6\exp(2x)$ $\qquad\qquad$ $[y = \exp(x)]$
7) $\quad 4y' + 7y + 2 = 2 + 23\exp(4x) - 11\exp(x)$ $\qquad$ $[y = \exp(4x) - \exp(x)]$

8) $\quad \overset{x}{\underset{0}{\circ}}y\,dx + 3\overset{x}{\underset{0}{\circ}}y\,dx = 16 + 6x^2$ $\qquad\qquad\qquad$ $[y = 4 + 3x]$

9) $\quad \overset{x}{\underset{0}{\circ}}\overset{x}{\underset{0}{\circ}}y^2 dx + \overset{x}{\underset{0}{\circ}}y\,dx = 2x + \frac{7}{2}x^2 + 2x^3 + \frac{3}{4}x^4$ $\qquad$ $[y = 3x + 2]$

10) $\quad y'' + \left(\frac{dy}{dx}\right)^2 + \overset{x}{\underset{0}{\circ}}y^2 dx = 13 + 28x + 22x^2 + \frac{17}{3}x^3 + 3x^4 + \frac{4}{5}x^5$ $\quad$ $[y = 2 + 3x + 2x^2]$

Differentiate the following: $\qquad\qquad\qquad\qquad\qquad\qquad\qquad$ ANSWERS
11) $\quad z = \dfrac{4x+2}{2x^2+2x}$ $\qquad\qquad\qquad\qquad\qquad\qquad$ $\dfrac{dz}{dx} = \dfrac{-1-2x-2x^2}{x^2+2x^3+x^4}$

12) $z = \dfrac{3+6x+9x^2+18x^3}{1+2x+4x^2+8x^3}$

$\dfrac{dz}{dx} = \dfrac{-6x-24x^2-24x^3}{(1+2x+4x^2+8x^3)^2}$

13) $z = \dfrac{2+7x+3x^2}{4+9x+2x^2}$

$\dfrac{dz}{dx} = \dfrac{10+16x+13x^2}{(4+9x+2x^2)^2}$

14) $z = \dfrac{1+4x+9x^2+3x^3}{2+3x+4x^2+9x^3}$

$\dfrac{dz}{dx} = \dfrac{5+28x+2x^2-54x^3-69x^4}{(2+3x+4x^2+9x^3)^2}$

15) $z = \dfrac{2+3x+4x^2+x^3}{2+4x+9x^2+2x^3}$

$\dfrac{dz}{dx} = \dfrac{-2-20x-17x^2-4x^3+x^4}{(2+4x+9x^2+2x^3)^2}$

Chapter 16

THE SOLUTION OF LINEAR AND NON-LINEAR PARTIAL DIFFERENTIAL EQUATIONS

In Chapter 15, the series method of solving ordinary differential equations was shown to become very powerful on adopting the vedic standpoint.

The same approach can be extended to partial differential equations, the assumed series solution now becoming two- or multi-dimensional. And just as procedures for expanding functions of polynomials proved invaluable in the previous chapter, so methods for expanding functions of bipolynomials are of considerable utility here.

We begin with a linear example.
The notation used is that:

(i) U_x & $U_{xx} = U_{x^2}$ denote respectively first and second order partial differentials of $U = U(x,y)$ with respect to x;

(ii) $U = U(a,b)$ denotes $U(x,y)$ evaluated with x=a & y=b, and $U_x(a,b)$ denotes the value of U_x at (a,b);

(iii) $U_{xx} = U_{x^2}$ & $U_{xxx} = U_{x^3}$, etc.

Example 1 Solve $U_{xx} + U_{yy} = 8x^2$ ...(1)

given that $U(0,0) = 2$, and $U_x(0,0) = U_{xx}(0,0) = U_{x^3}(0,0) \ldots = 0$,

and $U_y(0,0) = 0$, $U_{yx}(0,0) = 3$ & $U_{yxx}(0,0) = U_{yx^3}(0,0) = U_{yx^4}(0,0) = \ldots = 0$.

The complete working and solution can be written down as follows:

Assume that,

$$U = \overset{2}{a_0} + \overset{0}{b_0}x + \overset{0}{c_0}x^2 + \overset{0}{d_0}x^3 + \overset{0}{e_0}x^4 + \overset{0}{f_0}x^5 + \ldots$$

$$+ \overset{0}{a_1}y + \overset{3}{b_1}xy + \overset{0}{c_1}x^2y + \overset{0}{d_1}x^3y + \overset{0}{e_1}x^4y + \overset{0}{f_1}x^5y + \ldots$$

$$+ \overset{0}{a_2}y^2 + \overset{0}{b_2}xy^2 + \overset{4}{c_2}x^2y^2 + \overset{0}{d_2}x^3y^2 + \overset{0}{e_2}x^4y^2 + \overset{0}{f_2}x^5y^2 + \ldots$$

$$+ \overset{0}{a_3}y^3 + \overset{0}{b_3}xy^3 + \overset{0}{c_3}x^2y^3 + \overset{0}{d_3}x^3y^3 + \overset{0}{e_3}x^4y^3 + \overset{0}{f_3}x^5y^3 + \ldots$$

$$+ \overset{-\frac{2}{3}}{a_4}y^4 + \overset{0}{b_4}xy^4 + \overset{0}{c_4}x^2y^4 + \overset{0}{d_4}x^3y^4 + \ldots$$

$$+ \ldots \ldots$$

...(2)

Solution: $\underline{U = 2 + 3xy + 4x^2y^2 - \frac{2}{3}y^4}$

Explanation:
Since $U(0,0) = 2$, Equation (2) gives $a_0 = 2$
The other boundary conditions likewise give: $b_0 = 0$, $c_0 = d_0 = e_0 = \ldots = 0$,
$a_1 = 0$, $b_1 = 3$, $c_1 = d_1 = e_1 = \ldots = 0$

THIRD ROW:
Equating coefficients, firstly of unity, then of x, then of x^2 etc., a simple pattern emerges, on using Equation (2), with (1) substituted on the left-hand side:

Coefficient of unity $U_{xx} + U_{yy}$ gives $2c_0 + 2a_2 = 0$.
     Since $c_0 = 0$, $a_2 = 0$

Coefficient of x $U_{xx} + U_{yy}$ gives $6d_0 + 2b_2 = 0$.
     Since $d_0 = 0$, $b_2 = 0$

Coefficient of x^2 Here the term at the right-hand side of Equation (1) i.e. $8x^2$, enters the picture,
     $U_{xx} + U_{yy}$ gives $12e_0 + 2c_2 = 8$.
     Since $e_0 = 0$, $c_2 = 4$

Coefficient of x^3 $U_{xx} + U_{yy} = 20f_0 + 2d_2 = 0$.
     Since $f_0 = 0$, $d_2 = 0$
     Similarly, $e_2 = f_2 = \ldots = 0$

FOURTH ROW:
Coefficient of y $U_{xx} + U_{yy}$ gives $2c_1 + 6a_3 = 0$.
     Since $c_1 = 0$, $a_3 = 0$
     Similarly, $b_3 = c_3 = d_3 = \ldots = 0$

FIFTH ROW:

<u>Coefficient of y^2</u> $U_{xx} + U_{yy}$ gives $2c_2 + 12a_4 = 0$

$$8 + 12a_4 = 0$$

$$a_4 = -\tfrac{2}{3}$$

Continuing in this way, all other coefficients are found to be zero.

The second example uses the procedure for squaring bipolynomials outlined in Chapter 14.

Example 2   Solve $U^2 - 2xU_y = 25 - 10xy + 7x^2y^2 + 2x^2 - 10xy^3 + x^2y^6$ ...(3)

given $U(0,0) = 5$, & $U_x(0,0) = 0$, $U_{xx}(0,0) = U_{x^3}(0,0) = U_{x^4}(0,0) = \ldots..0$

The working and solution can be recorded thus:

$$
\begin{aligned}
U = {}& a_0 + b_0x + c_0x^2 + d_0x^3 + e_0x^4 + f_0x^5 + \ldots \\
& + a_1y + b_1xy + c_1x^2y + d_1x^3y + e_1x^4y + f_1x^5y + \ldots \\
& + a_2y^2 + b_2xy^2 + c_2x^2y^2 + d_2x^3y^2 + e_2x^4y^2 + f_2x^5y^2 + \ldots \\
& + a_3y^3 + b_3xy^3 + c_3x^2y^3 + d_3x^3y^3 + e_3x^4y^3 + f_3x^5y^3 + \ldots \\
& + a_4y^4 + b_4xy^4 + c_4x^2y^4 + d_4x^3y^4 + \ldots \\
& + \ldots\ldots
\end{aligned}
\quad\Bigg\}\quad ...(4)
$$

Solution: <u>$U = 5 - xy - xy^3$ exactly</u>

Explanation
Boundary conditions give: $a_0 = 5$, $b_0 = 0$, $c_0 = d_0 = e_0 = \ldots = 0$

ROW 2:
Substituting (4) in (3) and equating coefficients we have:

<u>For the coefficient of x</u>
$U^2 - 2xU_y$ gives $2a_0b_0 - 2a_1 = 0$
i.e. $0 - 2a_1 = 0$
$\therefore a_1 = 0$

<u>For the coefficient of x^2</u>
$U^2 - 2xU_y$ gives, $(2a_0c_0 + b_0^2) - 2b_1 = 2$
$\therefore b_1 = -1$

For the coefficient of x^3
$U^2 - 2xU_y$ gives, $(2a_0d_0 + 2b_0c_0) - 2c_1 = 0$
$\therefore\ c_1 = 0$
Similarly, $d_1 = e_1 = f_1 = \ldots = 0$

ROW 3:
For the coefficient of xy
$(2a_0b_1 + 2b_0a_1) - 4a_2 = -10$
$\therefore\ a_2 = 0$

For the coefficient of x^2y
$(2a_0c_1 + 2b_0b_1 + 2c_0a_1) - 4b_2 = 0$
$\therefore\ b_2 = 0$
Similarly, $c_2 = d_2 = e_2 = \ldots = 0$

ROW 4:
For the coefficient of xy^2
$(2a_0b_2 + 2a_1b_1 + 2a_2b_0) - 6a_3 = 0$
$\therefore\qquad\qquad 0 - 6a_3 = 0$
$\qquad\qquad\qquad a_3 = 0$

For the coefficient of x^2y^2
$(2a_0c_0 + 2b_0b_2 + 2c_0a_2 + 2a_1c_1 + b_1{}^2) - 6b_3 = 1$
$\therefore\qquad\qquad\qquad\qquad 1 - 6b_3 = 7$
$\qquad\qquad\qquad\qquad\qquad b_3 = -1$

For the coefficient of x^3y^2
$(2a_0d_2 + 2a_2d_0 + 2b_2c_0 + 2b_0c_2 + 2b_1c_1 + 2a_1d_1) - 6c_3 = 0$
$\therefore\qquad\qquad\qquad\qquad\qquad\qquad c_3 = 0$
Similarly, $d_3 = e_3 = f_3 = \ldots = 0$

ROW 5:
For the coefficient of xy^3
Ignoring the contributions to U^2 which vanish, we have:
$-8a_4 = 0$
$\therefore\ a_4 = 0$
Similarly, the remaining coefficients are found to be zero.

The next example uses the procedure for square-rooting bipolynomials.

Example 3 Solve $U^{1/2} + U_y = 1 + 2x$
$\qquad\qquad\qquad\qquad + xy + 2x^2y$ ...(5)

given $U(0,0) = 1$, & $U_x(0,0) = U_{xx}(0,0) = U_{x^3}(0,0) = \ldots = 0$

The working of the first solution can be written down thus:

$$\begin{aligned}
U = \overset{1}{a_0} + \overset{0}{b_0 x} + \overset{0}{c_0 x^2} + \overset{0}{d_0 x^3} + \overset{0}{e_0 x^4} + \overset{0}{f_0 x^5} + \ldots \\
+ \overset{0}{a_1 y} + \overset{2}{b_1 xy} + \overset{0}{c_1 x^2 y} + \overset{0}{d_1 x^3 y} + \overset{0}{e_1 x^4 y} + \overset{0}{f_1 x^5 y} + \ldots \\
+ \overset{0}{a_2 y^2} + \overset{0}{b_2 xy^2} + \overset{1}{c_2 x^2 y^2} + \overset{0}{d_2 x^3 y^2} + \overset{\sigma}{e_2 x^4 y^2} + \overset{0}{f_2 x^5 y^2} + \ldots \\
+ \overset{0}{a_3 y^3} + \overset{0}{b_3 xy^3} + \overset{0}{c_3 x^2 y^3} + \overset{0}{d_3 x^3 y^3} + \overset{0}{e_3 x^4 y^3} + \overset{0}{f_3 x^5 y^3} + \ldots \\
+ \overset{0}{a_4 y^4} + \overset{0}{b_4 xy^4} + \overset{0}{c_4 x^2 y^4} + \overset{0}{d_4 x^3 y^4} + \ldots \\
+ \ldots \ldots
\end{aligned}$$
...(6)

$$\begin{aligned}
U^{\frac{1}{2}} = \overset{1}{A_0} + \overset{0}{B_0 x} + \overset{0}{C_0 x^2} + \overset{0}{D_0 x^3} + \overset{0}{E_0 x^4} + \overset{0}{F_0 x^5} + \ldots \\
+ \overset{0}{A_1 y} + \overset{1}{B_1 xy} + \overset{0}{C_1 x^2 y} + \overset{0}{D_1 x^3 y} + \overset{0}{E_1 x^4 y} + \overset{0}{F_1 x^5 y} + \ldots \\
+ \overset{0}{A_2 y^2} + \overset{0}{B_2 xy^2} + \overset{0}{C_2 x^2 y^2} + \overset{0}{D_2 x^3 y^2} + \overset{0}{E_2 x^4 y^2} + \overset{0}{F_2 x^5 y^2} + \ldots \\
+ \overset{0}{A_3 y^3} + \overset{0}{B_3 xy^3} + \overset{0}{C_3 x^2 y^3} + \overset{0}{D_3 x^3 y^3} + \overset{0}{E_3 x^4 y^3} + \overset{0}{F_3 x^5 y^3} + \ldots \\
+ \ldots \ldots
\end{aligned}$$
...(7)

i.e. the first solution is $\underline{U = 1 + 2xy + x^2 y^2}$

Explanation

FIRST ROW OF U

From boundary conditions, taking the square root of 1 to be +1, to obtain a first solution, we have:

$a_0 = +1$, and $b_0 = c_0 = d_0 = \ldots . = 0$

FIRST ROW OF $U^{\frac{1}{2}}$

The first row of coefficients of $U^{\frac{1}{2}}$ is worked out using the procedure for square-rooting the first row of a bipolynomial, given in Chapter 14—or else directly from the observation that $1^{\frac{1}{2}} = \pm 1$. Hence $A_0 = +1$ (first solution) and $B_0 = C_0 = D_0 = \ldots = 0$

SECOND ROW OF U

Here we successively equate coefficients of 1, x, x^2, x^3, etc. It is convenient to put down the contributions of the terms $U^{\frac{1}{2}}$, U_y, and the right-hand side of Equation (5), in a table. Note that in each row of the table the fourth column equals the sum of the second and third columns, corresponding to Equation (5).

Coefficient of	$U^{\frac{1}{2}}$	U_y	r.h.s. of Equation (5)	Coefficient evaluated from Equation (5)
1	1	a_1	1	$a_1 = 1 - 1 = 0$
x	0	b_1	2	$b_1 = 2$
x^2	0	c_1	0	$c_1 = 0$
x^3	0	d_1	0	$d_1 = 0$

Similarly, $e_1 = f_1 = \ldots = 0$

We can now use the second row of coefficients of U to continue the square-rooting process, and find a second row of coefficients for $U^{\frac{1}{2}}$:

$A_1 = a_1 \div 2A_0 = 0$
$B_1 = (b_1 - 2A_1B_0) \div 2A_0$
$\quad = (2 - 0) \div 2 = 1$
$C_1 = (c_1 - 2A_1C_0 - 2B_1B_0) \div 2A_0$
$\quad = (0 - 0 - 0) \div 2 = 0$
$D_1 = (d_1 - 2A_1D_0 - 2B_1C_0 - 2C_1B_0) \div 2A_0$
$\quad = 0$

Similarly, $E_1 = F_1 = \ldots = 0$

THIRD ROW OF U

Coefficient of	$U^{\frac{1}{2}}$	U_y	r.h.s. of Equation (5)	Coefficient evaluated from Equation (5)
y	0	$2a_2$	0	$2a_2 = 0$
xy	1	$2b_2$	1	$2b_2 = 0$
x^2y	0	$2c_2$	2	$2c_2 = 2$
x^3y	0	$2d_2$	0	$2d_2 = 0$

Similarly, $e_2 = f_2 = \ldots = 0$

THIRD ROW OF $U^{\frac{1}{2}}$
$\quad A_2 = (a_2 - A_1^2) \div 2A_0 = 0$
$\quad B_2 = (b_2 - 2A_2B_0 - 2A_1B_1) \div 2A_0 = 0$
$\quad C_2 = c_2 - 2A_1C_1 - 2A_2C_0 - 2B_2B_0 - B_1^2$
$\quad\quad = 1 - 1^2 = 0$

Similarly, $D_2 = E_2 = F_2 = \ldots = 0$

FOURTH ROW OF U

Coefficient of	$U^{\frac{1}{2}}$	U_y	r.h.s. of Equation (5)	Coefficient evaluated
y^2	0	$3a_3$	0	$a_3 = 0$
xy^2	0	$3b_3$	0	$b_3 = 0$
x^2y^2	0	$3c_3$	0	$c_3 = 0$

Likewise, $d_3 = e_3 = f_3 = \ldots = 0$
The remaining coefficients of U and $U^{1/2}$ are now found to be zero.

A second solution arises from using the other square root of $a_0 = 1$, namely -1. Taking $A_0 = -1$ gives the solution:

$$
\begin{aligned}
U = {} & \overset{1}{a_0} + \overset{0}{b_0 x} + \overset{0}{c_0 x^2} + \overset{0}{d_0 x^3} + \overset{0}{e_0 x^4} + \ldots \\
& + \overset{2}{a_1 y} + \overset{2}{b_1 xy} + \overset{0}{c_1 x^2 y} + \overset{0}{d_1 x^3 y} + \overset{0}{e_1 x^4 y} + \ldots \\
& + \overset{\frac{1}{2}}{a_2 y^2} + \overset{1}{b_2 xy^2} + \overset{1}{c_2 x^2 y^2} + \overset{0}{d_2 x^3 y^2} + \overset{0}{e_2 x^4 y^2} + \ldots \\
& + \overset{-\frac{1}{12}}{a_3 y^3} + \overset{-\frac{1}{6}}{b_3 xy^3} + \overset{\frac{1}{6}}{c_3 x^2 y^3} + \overset{0}{d_3 x^3 y^3} + \overset{0}{e_3 x^4 y^3} + \ldots \\
& + \overset{\frac{5}{96}}{a_4 y^4} + \ldots \\
& + \ldots\ldots
\end{aligned}
$$

$$
\begin{aligned}
U^{1/2} = {} & \overset{-1}{A_0} + \overset{0}{B_0 x} + \overset{0}{C_0 x^2} + \overset{0}{D_0 x^3} + \overset{0}{E_0 x^4} + \ldots \\
& + \overset{-1}{A_1 y} + \overset{-1}{B_1 xy} + \overset{0}{C_1 x^2 y} + \overset{0}{D_1 x^3 y} + \overset{0}{E_1 x^4 y} + \ldots \\
& + \overset{\frac{1}{4}}{A_2 y^2} + \overset{\frac{1}{2}}{B_2 xy^2} + \overset{-\frac{1}{2}}{C_2 x^2 y^2} + \overset{0}{D_2 x^3 y^2} + \overset{0}{E_2 x^4 y^2} + \ldots \\
& + \overset{-\frac{5}{24}}{A_3 y^3} + B_3 xy^3 + C_3 x^2 y^3 + D_3 x^3 y^3 + E_3 x^4 y^3 + \ldots \\
& + \ldots\ldots
\end{aligned}
$$

Example 4 Solve $U + \ln(1 + 2x^2 + xU_x) = 2 - x + x^2 + y$ ...(8)

where $U = U(x,y)$

Assume that $\quad U = a_0 + b_0 x + c_0 x^2 + d_0 x^3 + e_0 x^4 + \ldots$

$$
\left.
\begin{aligned}
& + a_1 y + b_1 xy + c_1 x^2 y + d_1 x^3 y + e_1 x^4 y + \ldots \\
& + a_2 y^2 + b_2 xy^2 + c_2 x^2 y^2 + d_2 x^3 y^2 + \ldots \\
& + a_3 y^3 + b_3 xy^3 + c_3 x^2 y^3 + \ldots \\
& + \ldots
\end{aligned}
\right\} \quad \ldots(9)
$$

Then from (9), $1 + 2x^2 + xU_x = 1 + b_0 x + (2c_0 + 2)x^2 + 3d_0 x^3 + 4e_0 x^4 + \ldots$
$$
\begin{aligned}
& + b_1 xy + 2c_1 x^2 y + 3d_1 x^3 y + \ldots \\
& + b_2 xy^2 + 2c_2 x^2 y^2 + 3d_2 x^3 y^2 + \ldots \\
& + \ldots \hspace{8cm} \ldots(10)
\end{aligned}
$$

Substituting Expression (9) in Equation (8), and differentiating partially w.r.t. 'x', and multiplying up by $1 + 2x^2 + xU_x$, (using (10)), we have, on equating coefficients:

$$
\begin{pmatrix}
\overset{-\frac{1}{2}}{b_0} & \overset{-\frac{7}{12}}{+2c_0x} & \overset{-\frac{1}{2}}{+3d_0x^2} & +4e_0x^3 + \dots \\
+b_1y & +2c_1xy & +3d_1x^2y & +4e_1x^3y + \dots \\
+b_2y^2 & +2c_2xy^2 & +3d_2x^2y^2 & +4e_2x^3y^2 \\
+b_3y^3 & +\dots & &
\end{pmatrix}
\begin{pmatrix}
1 & \overset{-\frac{1}{2}}{+b_0x} & \overset{\frac{17}{12}}{+(2c_0+2)x^2} & \overset{-\frac{1}{2}}{+3d_0x^3 + \dots} \\
& b_1xy & +2c_1x^2y & +3d_1x^3y + \dots \\
& +b_2xy^2 & +2c_2x^2y^2 & +\dots \\
& +b_3xy^3 & +\dots &
\end{pmatrix}
$$

$$
+
\begin{pmatrix}
\overset{-\frac{1}{2}}{b_0} & \overset{\frac{17}{6}}{+(4c_0+4)x} & \overset{-\frac{3}{2}}{+9d_0x^2} & +16e_0x^3 \\
+b_1y & +4c_1xy & +9d_1x^2y & +16e_1x^3y \\
+b_2y^2 & +4c_2xy^2 & +9d_2x^2y^2 & +\dots \\
+b_3y^3 & +\dots & &
\end{pmatrix}
$$

$$
= (-1+2x)
\begin{pmatrix}
\overset{1}{1} & \overset{-\frac{1}{2}}{+b_0x} & \overset{\frac{17}{12}}{+(2c_0+2)x^2} & \overset{-\frac{1}{2}}{+3d_0x^3} & +4e_0x^4 + \dots \\
& +b_1xy & +2c_1x^2y & +\dots &
\end{pmatrix}
\qquad \dots(11)
$$

The equating of coefficients requires also use of the next equation.
Substituting Expression (9) in Equation (8), and differentiating partially w.r.t. 'y', and multiplying up by $1 + 2x^2 + xU_x$, gives Equation (12):

$$
\begin{pmatrix}
\overset{0}{(a_1-1)} & \overset{0}{+b_1x} & \overset{0}{+c_1x^2} & +d_1x^3 + \dots \\
+2a_2y & +2b_2xy & +2c_2xy^2 & +\dots \\
+3a_3y^2 & +3b_3xy^2 & +\dots & \\
+4a_4y^3 & +\dots & & \\
+\dots & & &
\end{pmatrix}
\begin{pmatrix}
1 & \overset{-\frac{1}{2}}{+b_0x} & \overset{\frac{17}{12}}{+(2c_0+2)x^2} & \overset{-\frac{1}{2}}{+3d_0x^3 + \dots} \\
+0y & b_1xy & +2c_1x^2y & +\dots \\
+0y^2 & +b_2xy^2 & +\dots & \\
+0y^3 & +\dots & &
\end{pmatrix}
$$

$$
+
\begin{pmatrix}
\overset{0}{0} & \overset{0}{+b_1x} & \overset{0}{+2c_1x^2} & +3d_1x^3 & +4e_1x^4 + \dots \\
0y & +2b_2xy & +4c_2x^2y & +6d_2x^3y & +\dots \\
0y^2 & +3b_3xy^2 & +\dots & &
\end{pmatrix} = 0
\qquad \dots(12)
$$

Since the partial derivative w.r.t. 'y' of the right-hand side of Equation (8) is 1, which is then multiplied by $1 + 2x^2 + xU_x$, the resulting term has been taken to the left-hand side of Equation (11). It shows up, there, by changing the a_1 to a_1-1, in the expression for U_y.

The steps by which coefficients are obtained are as follows:

Put x = y = 0 in Equations (8) and (9) to obtain $a_0 = 2$.
Thereafter:

Equation	Coefficient of	Coefficient evaluated	
(11)	1	$2b_0$	$= -1$
(11)	x	$(2+4)c_0$	$= -\frac{7}{4}$
(11)	x^2	$(3+9)d_0$	$= -2$
(11)	x^3	$20e_0$	$= \frac{527}{144}$
(12)	1	a_1	$= 1$
(12)	x	$2b_1$	$= 0$
(12)	x^2	$3c_1$	$= 0$
(12)	x^3	$4d_1$	$= 0$
	Similarly, $e_1 = f_1 = g_1 = \ldots = 0$		
(12)	y	$2a_2$	$= 0$
(12)	xy	$4b_2$	$= 0$
	Similarly, $c_2 = d_2 = e_2 = \ldots = 0$		
	& $a_3 = b_3 = c_3 = \ldots = 0$		

☐ the solution is: $\underline{U = y + 2 - \frac{1}{2}x - \frac{7}{24}x^2 - \frac{1}{6}x^3 + \frac{527}{2880}x^4 + \ldots}$

Exercise

Solve the following equations:

1) $U_{xx} + U_{yy} = 8 + 18x + 6y$

given that $U(0,0) = 2, U_{x^4}(0,0) = U_{x^2}(0,0) = \ldots = 0$
$U_y(0,0) = 0, U_{yx}(0,0) = 2, U_{yx^2}(0,0) = \ldots = 0$

$[U = 2 + 4x^2 + 3x^3 + 2xy + 3x^2y]$

2) $U_x + 6y + U_y = 5x + 9y + 2$

given $U(0,0) = 4, U_x(0,0) = 2 = U_{xx}(0,0), U_{x^3}(0,0) = \ldots = 0$
$U_y(0,0) = 0, U_{yx}(0,0) = 3, U_{yx^2}(0,0) = \ldots = 0$

$[U = x^2 + 3xy + 2x + 4]$

3) $U_{x^2} + 6xU_y + 3y = 4$

given $U(0,0) = 3, U_x(0,0) = 1, U_{x^2}(0,0) = \ldots = 0$
$U_y(0,0) = U_{yx}(0,0) = U_{yx^2}(0,0) = \ldots = 0$

$[U = 2x^2 + x + 3]$

4) $U_{x^3} + U_{y^2} + U_{xy} = 12$

given $U(0,0) = 4, U_x(0,0) = 3, U_{x^2}(0,0) = 4, U_{x^3}(0,0) = 6$
$U_{x^4}(0,0) = \ldots = 0, U_y(0,0) = 2, U_{y^2}(0,0) = 2,$
$U_{y^3}(0,0) = \ldots = 0. U_{xy}(0,0) = 4, U_{x^2y}(0,0) = \ldots = 0$

$[U = x^3 + 2x^2 + 3x + 4 + y^2 + 2y + 4xy]$

5) $U_{x^2} + U_{y^2} = 16$

given $U(0,0) = 8$, $U_y(0,0) = 6$, $U_{yy}(0,0) = 4$
$U_{y^3}(0,0) = ... = 0$. $U_{xy}(0,0) = 3$,

$U_{x^2y}(0,0) = ... = 0$, $U_x(0,0) = 4$, $U_{x^2}(0,0) = 12$, $U_{x^3}(0,0) = ... = 0$

$[U = 6x^2 + 4x + 8 + 2y^2 + 6y + 3xy]$

6) $U^{1/2} + 2xU_y = 16x^2 + 26x + 16xy + 2y + 3$

given $U(0,0) = 9$, $U_x(0,0) = 12$, $U_{x^2}(0,0) = 8$
$U_{x^3}(0,0) = ... = 0$, $U_y(0,0) = 12$, $U_{y^2}(0,0) = 8$
$U_{y^3}(0,0) = ... = 0$, $U_{xy}(0,0) = 8$,
$U_{x^2y}(0,0) = ... = 0$

$[U = 4x^2 + 12x + 4y^2 + 12y + 8xy + 9]$

7) $yU_{x^2} + xU_{y^2} + U_{xy} - 2U^{\frac{1}{2}} = 0$

given $U(0,0) = 1$, $U_x(0,0) = 2$, $U_{x^2}(0,0) = 2$
$U_{x^3}(0,0) = ... = 0$, $U_y(0,0) = 2$, $U_{y^2}(0,0) = 2$
$U_{y^3}(0,0) = ... = 0$, $U_{xy}(0,0) = 2$
$U_{x^2y}(0,0) = ... = 0$

$[U = x^2 + 2x + y^2 + 2y + 2xy + 1]$

8) $U^{1/2} - U_{x^3} + 2U_{xy} = x^2 - 22x + 3y$

given $U(0,0) = 0 = U_x(0,0)$, $U_{x^2}(0,0) = 8$, $U_{x^3}(0,0) = 24$,
$U_{x^4}(0,0) = 24$, $U_{x^5}(0,0) = ... = 0$, $U_{y^2}(0,0) = 18$,
$U_y(0,0) = U_{y^3}(0,0) = ... = 0$. $U_{xy}(0,0) = 12$, $U_{x^2y}(0,0) = 6$,
$U_{x^3y}(0,0) = ... = 0$

$[U = x^4 + 4x^3 + 6x^2y + 4x^2 + 12xy + 9y^2]$

9) $U_x - 6U^{1/2} = 0$

given $U_{x^2}(0,0) = 18$, $U(0,0) = U_x(0,0) = U_{x^3}(0,0) = ... = 0$,
$U_{y^2}(0,0) = 32$, $U_y(0,0) = U_{y^3}(0,0) = ... = 0$,
$U_{xy}(0,0) = 24$, $U_{x^2y}(0,0) = ... = 0$

$[U = 9x^2 + 24xy + 16y^2]$

10) $U^{1/2} + U^2 - U_{x^2y^2} = x^4y^4 + xy - 1$

given $U_{x^3y^2}(0,0) = 1$, $U(0,0) = U_x(0,0) = U_{x^2}(0,0) = ... = 0$,
$U_y(0,0) = U_{y^2}(0,0) = ... = 0$, $U_{xy}(0,0) = U_{xy^2}(0,0) = ... = 0$
$U_{yx^2}(0,0) = U_{yx^3}(0,0) = ... = 0$

$[U = x^2y^2]$

C3